Murat Özyüksel is Professor of History and Political Science at Istanbul University.

THE HEJAZ RAILWAY AND THE OTTOMAN EMPIRE

Modernity, Industrialisation and
Ottoman Decline

MURAT ÖZYÜKSEL

I.B. TAURIS

LONDON · NEW YORK

Published in 2014 by I.B.Tauris & Co. Ltd
6 Salem Road, London W2 4BU
175 Fifth Avenue, New York NY 10010
www.ibtauris.com

Distributed in the United States and Canada
Exclusively by Palgrave Macmillan
175 Fifth Avenue, New York NY 10010

Library of Ottoman Studies, 39

ISBN: 978 1 78076 364 4
eISBN: 978 0 85773 743 4

A full CIP record for this book is available from the British Library
A full CIP record is available from the Library of Congress

Library of Congress Catalog Card Number: available

Typeset in Garamond Three by OKS Prepress Services, Chennai, India
Printed and bound by CPI Group (UK) Ltd, Croydon, CR0 4YY

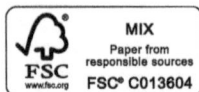

Translated from Turkish by Sezin Tekin

TABLE OF CONTENTS

LIST OF ILLUSTRATIONS

Source for all images: İstanbul Üniversitesi Nadir Eserler Kütüphanesi

ABBREVIATIONS

Admin.Reg	Administrative Registratur
AfEW	Archiv für Eisenbahnwesen
B.D.	British Documents on the Origins of the War, 1898–1914
B.D.F.A.	British Documents on Foreign Affairs
B.O.A	Başbakanlık Osmanlı Arşivi
ÇYO	Çağını Yakalayan Osmanlı
DHP	Société Ottomane du Chemin de fer de Damas-Hamah et Prolongements
DLZ	Deutsche Levante Zeitung
G.D.D.	German Diplomatic Documents, 1871–1914
G.P.	Die Grosse Politik der Europäischen Kabinette
HHStA	Haus-, Hof-, und Staatsarchiv
PA/AA	Politisches Archiv des Auswaertigen Amtes
P.D.H.C.	Parliamentary Debates, House of Commons
ZdVDEV	Zeitung des Vereins Deutscher Eisenbahnverwaltungen

MAP

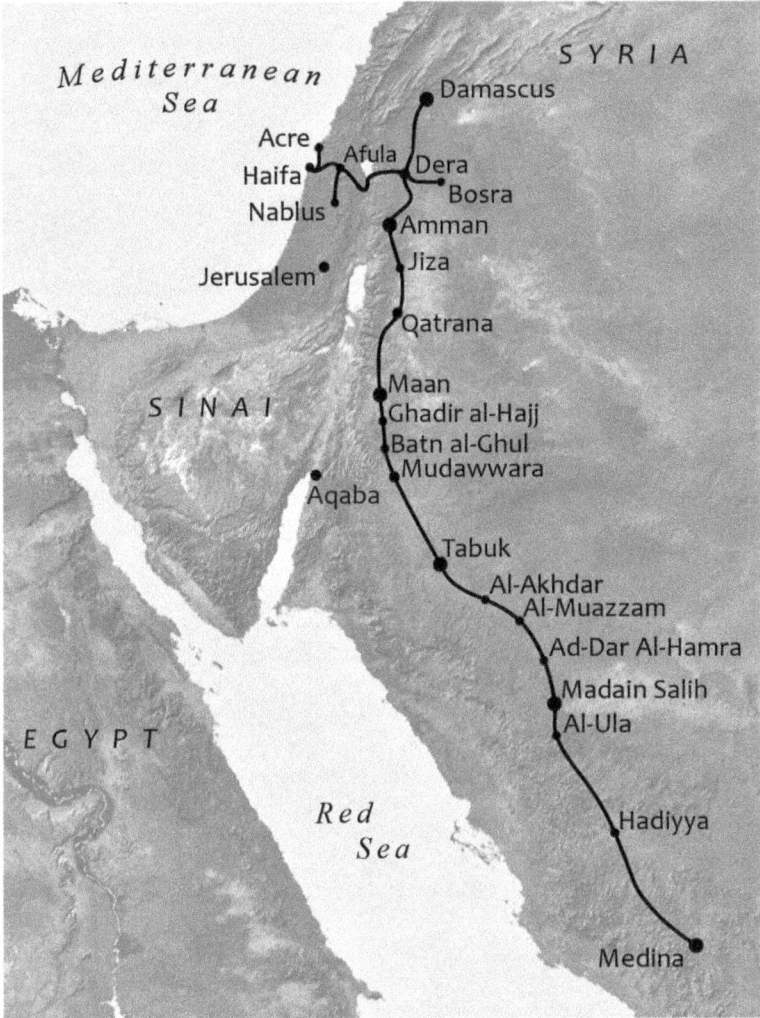

Route of the Hejaz Railway

INTRODUCTION

God bless the man wi'peace and plenty
That first invented metal plates.
Draw out his days to five times twenty,
Then slide him through the heavenly gates.[1]

Very few inventions have influenced the destiny of humankind to
the extent that railways have. There can be no doubt that railways
have brought about radical change in the economic, political, social
and intellectual structures of many societies. While the politics and
the ideology of the nineteenth century were determined by the
French Revolution, the economic history of the era was determined
largely by the British industrial revolution. One of the important
dynamics in this process was the tremendous progress achieved in
transportation thanks to railways. Railways both introduced the
meta-production process to rural areas and contributed extensively
to the solution of problems related to labor, raw material and
markets. Industrial products manufactured by means of steam
engines could be swiftly delivered to far-off markets by
locomotives, and raw materials could also be dispatched to
industrial hubs. Moreover, railways facilitated the flow of labor
from rural areas to urban areas, thus contributing to the solution of

the manpower shortage. The function of the railways was significant not only in that they connected the rural areas of an industrial country to its own market, but also because they tied under-developed countries into this chain. Put differently, railways were one of the major determining factors in the formation of the capitalist world economy.

The main energy resource of the eighteenth century was coal. The cost of transporting coal from mines to points of consumption sometimes well exceeded the cost of production. The primary reason behind the growth of British mining was the proximity of the coal mines to the sea. In 1825, an engineer named Stephenson constructed a railway to connect the Darlington mine to the Port of Stockton. Coal prices in Stockton promptly dropped by half. Realizing the significance of this novel vehicle, some businessmen decided to finance a railway project between Manchester and Liverpool; the first modern railway was inaugurated on the aforementioned line on September 15, 1830. Construction proceeded at an astounding pace: by 1852, there remained only three cities in the UK not connected to a railway line.

Locomotives began to operate in 1832 in France, in 1835 in Germany, and in 1836 in Belgium. In the course of a few years, train journeys commenced in Austria, Russia, the Netherlands and Italy. In 1843, the first international railway line began to operate between Liège and Cologne. Meanwhile, the American continent was in no way lagging behind. In the years 1851–52 railway lines were inaugurated in Chile, Peru, Brazil and Argentina. In order to grasp the speed of growth, it should suffice to say that the length of railways lines, which had totaled around 5,000 miles in 1840, reached 500,000 miles in the year 1900. Rails, bridges, viaducts and tunnels had transformed the landscape of the earth.

The renowned historian Hobsbawm says, "No innovation of the Industrial Revolution fired the imagination of humankind as much as the railway, as witness the fact that it is the only product of nineteenth-century industrialization which has been fully absorbed into the imagery of popular and literate poetry."[2] Pommier wrote the lines below, as it were, to justify Hobsbawm's remarks:

Devant de tels témoins, o secte progressive
Vantez-nous le pouvoir de la locomotive
Vantez-nous le vapeur et les chemins de fer.[3]

Naturally, the rate of railway construction was not the same everywhere. The speed of this development was very high in industrializing countries, where railway construction was driven by the internal dynamics of their own economies. The more the railways gave momentum to industrialization, the greater the need for new lines to cater to the requirements of industry. In the first two decades of railways, iron and coal production in the UK tripled. This increase is understandable, since approximately 300 tons of iron per mile was required for the rails alone. As railway investments speeded up, it became clear that steel lines were much more durable than iron ones. Therefore, railways became one of the main factors leading to the establishment of the modern steel industry. Production of cheap and high-quality steel paved the way for the development of new industries such as shipbuilding, construction and the heavy chemical industry.

In addition to capital accumulation, railways contributed to the establishment of the new lifestyle called capitalism by creating new areas of employment and with them, needless to say, new problems. In 1847 around 50,000 workers were employed on existing lines, while 250,000 workers were constructing new lines. Thousands of others were producing iron, rails, construction material, locomotives and coal for these railways. By 1907 the number of people working only on the railways exceeded 600,000. For the first time in history, workers were working very intensively and under the same conditions as their peers. This resulted in the first labor movements. Factory smokestacks, clanging rails, and mushrooming workers' districts — all a harbinger of a new era that would be marked by political freedom.

Not only industrializing countries joined in the boom: even those nations whose level of economic development did not yet require railway construction could not seem to stay out of the process. For instance, the railways in India were constructed by the colonialists themselves. Railways opened the remotest corners of the Indian market to British capital. Interestingly enough, Britain, seeing that the railways they constructed encouraged the development of local industry, impeded this progress by implementing various economic mechanisms such as flexible tariffs. While the fees collected for goods carried between the ports and the centers of production were low, those collected from the lines which interconnected the centers in India were higher – so much so that transporting coal from Raniganj, which had rich coal mines, to neighboring states was more expensive than importing coal from Britain. As a result, while India's natural resources were being tied to the capitalist world economy, the country itself was prevented from rendering production in line with its own requirements.

In the nineteenth century, capital accumulation gained momentum, with export activities starting to take the form of capital investments. Capital, which was initially exported as external debt, later on shifted to mines, irrigation systems and particularly to railways outside the country. India was not, of course, a unique case. Britain embarked on railway investments in countries outside its borders, as well; namely, Belgium, Denmark, Austria, Canada and Argentina. Gradually German, French, Russian, and Belgian capital undertook railway investments in other countries. National governments supported railway investments not only because they facilitated economic penetration into an underdeveloped country, but also because they ensured that the region the lines passed through came under the influence of the country in question. Briefly stated, railway construction had become one of the major tools of economic and political expansionism. It should come as no surprise that the Ottoman Empire, which was one of the leading spheres of competition for the imperialist countries, should attract competing investments from numerous foreign companies, all jockeying for advantage. As a matter of fact, by the end of the nineteenth century, constructing railways on Ottoman territory had become the most effective way for the imperialist countries to gain influence there.

The structure of the Ottoman economy was not, however, dynamic enough to call for railway construction. Nonetheless, in the 1830s, during the earliest years of railways, Britain was engineering plans for prospective railways to be constructed in the Ottoman territory. The İzmir-Aydın, İzmir-Kasaba, Constanza-Cernavoda, and Varna-Ruse lines were later operated by the British in order to make use of the agricultural potential of these regions. It should be noted that the years during which the British constructed railways in the Ottoman territory coincided with the years in which the empire pursued a policy of territorial integrity. As is known, Britain would largely abandon this policy in the aftermath of the War of 1877–78. From then on, Germany would aspire to this role and try to implement a gigantic railway project that would start from İzmit and continue all the way to Baghdad. Germans undertook this project with the aim of increasing their influence on the Ottoman Empire and benefiting from the abundant potential of its territory. The strategic importance of the Anatolian and Baghdad railways and the desire to rely on the Ottoman army in case of war also played a role in this effort.

The Anatolian and Baghdad railways which the Germans tried to construct were the main topic of my previous work.[4] In the present work, on the other hand, I try to shed light on the most intriguing part of Ottoman railway history, namely the history of the Hejaz railway, the one and only railway line that the Ottoman Empire managed to construct and operate. My study of the Anatolian/Baghdad railways called for the analysis of a complicated process, involving not only Ottoman-German relations, but the empire's connections to Britain, France, Austria and Russia. In the same vein, a study of the Hejaz railway cannot disregard the power relations competing for Syria and the Hejaz. As a matter of fact, the Hejaz railway became the heart of competition among the Ottoman/German block, Britain and France. Needless to say, Bedouins, the amirs of Mecca, and urban Arabs, who had begun to adopt the ideology of nationalism, are local players to be included in this analysis. Under these circumstances, when the potential for conflict, predating the Great War, combined with the complex political, social and economic structures of the region, imperialist policies assumed a very interesting dimension. As this study makes clear, in order to attain their goals, imperialist states pursued changing policies of alliance in different regions and environments.

The Hejaz railway project came into being as part and parcel of Abdülhamid's absolutist/Islamist policy. As a result of losing the Balkans, the majority of the Christian community had seceded from the empire, leaving the Ottoman Empire to adopt a Muslim/Asian identity. Abdülhamid conceded the situation and focused his efforts on holding at least the Muslim elements of the empire together. However, Britain, starting to play its Arab card overtly, undermined the Islamist policy of the sultan, which was aimed at averting the spread of nationalism to Muslim peoples other than Turks. In this context, the Hejaz railway can be regarded as Abdülhamid's final struggle in the Arab territory within the framework of a pro-Islamist policy that he was trying to implement on extremely slippery ground.

Abdülhamid and the Ottoman rulers put tremendous effort into the completion of the Hejaz railway. Despite all their efforts and sacrifices, however, at a certain point they got stuck. Contrary to what was initially asserted by foreign observers, that point had nothing to do with financial resources or lack of technical staff. These problems, the Ottoman officials managed to overcome. For example, internal and external grants generated by local Muslim subjects as well as Muslims throughout the world constituted around one third of the funds. At the sultan/caliph's call, Muslims from all over the world contributed to the cost of the "sacred line". At the beginning, foreigners, primarily Germans, were employed as technical staff. Over time, however, the endeavor to employ Ottoman engineers proved to a certain extent successful. The employment of military personnel played a significant role in responding to the demand for manpower. Yet the solution of these problems could be likened to temporary periods of improvement in the health of a dying patient. Abdülhamid had accomplished the financially and technically challenging part of the project and the Ottomans seemed to be just one step away from their goal. But in the end they had to desperately abort the construction: the resistance of the Amir of Mecca and the Bedouins precluded the progress of the lines beyond Medina. These two forces, also backed by the British, were well aware that the Hejaz railway would take away some of their income sources, such as camel transportation, and put an end to their disorderly behavior. Moreover, upon the ultimatum of Britain, the Ottomans had not been able to complete the Aqaba exit of the Hejaz railway. Nor could they complete the Afula−Jerusalem line, as a result of pressure exerted by the French. The patient had been determined to recover (Abdülhamid decided to construct the Hejaz railway); some temporary periods of improvement were observed (financial and technical problems were overcome and most of the railway was constructed), but, since the whole body had failed miserably (given the economic dependency, military and political incapacity of the Ottoman Empire), the treatment did not work. Seen in this light, the Hejaz railway constituted a distressing part of the last years leading up to the collapse of the Ottoman Empire. It became impossible for the Ottoman Empire to compete with foreign railway companies on its own territory.

This study, which aims to cover this complex and interesting process in its various aspects, is based mainly on research in Ottoman, German and Austrian archives. In addition to these records, books, brochures, newspapers and other contemporary publications, and published British documents proved especially helpful. At first glance, it might sound astonishing that Germans displayed such a keen interest in the Hejaz railway and could obtain such accurate information about it. However, considering that the Germans regarded the Hejaz railway as an extension of the Baghdad railway and hence a good opportunity to penetrate deeper into the Ottoman Empire, this situation becomes more comprehensible. In a significant portion of the German documents, measures were set forth to protect the Hejaz line against pressures from the British and the French.

Chapter One gives general information regarding the historical development of railway construction in the Ottoman Empire, elaborating the internal and external dynamics of the process. Regions of influence that the British, Germans and French created thanks to the privileges they enjoyed on the railways are discussed in tandem with the welcoming attitudes of the Ottoman rulers. This section includes a deeper analysis of the competition between the British and French over railways in the region of Syria, a region of particular interest due to its close relevance to the Hejaz railway.

Chapter Two analyzes the main determining factors in the construction of the Hejaz railway. It highlights the fact that economic factors did not contribute at all to the decision-making process, while religious/political and military factors both proved decisive. In this regard, the Hejaz railway project is positioned within the framework of Abdülhamid's general pro-Islamist policy. In line with this analysis, I explore what pan-Islamism signified for Abdülhamid, the Germans and the European states, all of which had Muslim colonies. This section clarifies the significance of the holy cities and the pilgrimage for all powers related to the region, especially for Abdülhamid and Britain.

Chapter Three deals with the financing of the Hejaz railway. Around one-third of the project was financed through grants, which the Austrian Ambassador Pallavicini regarded as "a form of fund generation unique in economic history". The most interesting dimension is the grants given by Muslims living outside the Ottoman borders in the name of Islamic solidarity. However, it is emphasized that this dimension should not be overstated, as external grants constituted a negligible part of the total financing. This part also discusses at length the role of the press in raising grants for the Hejaz railway.

Chapter Four analyzes the construction process of the Hejaz railway. It includes a detailed discussion concerning the negotiations with the managers of DHP, a French company which was operating in the region. The direct involvement of the French Foreign Minister and Ambassador in the heart of the negotiations is an indicator that the issue was far more crucial than mere local trade relations. As a matter of fact, "national interest" discourse surpassed the commercial expectations of the company in the conclusion of the negotiations. The decisive factor in the trade competition between the Hejaz railway administration and the DHP was the preconditions of the Ottoman–French loan agreement, which were imposed by the French. The formation of the railway administration and the technical staff, and the difficulties that had to be overcome in the construction process are among the issues that are covered in this part. The employment of Ottoman engineers in the construction, the resolution of the manpower shortage by employing soldiers at the construction site and similar efforts are analyzed in this context.

Chapter Five discusses the mechanisms employed by the amirs of Mecca and the Bedouins, the British and the French to prevent the construction of the Hejaz railway. The reasons behind the radicalization of the resistance shown by the Bedouins and the amirs of Mecca against the extension of the line from Medina to Mecca and the construction of the Jeddah–Mecca line are also explored. The Aqaba incident was exemplary in that it demonstrated the limits to British tolerance of the Hejaz railway. Once the British felt that Egypt was under threat, they went as far as issuing an ultimatum. Furthermore, it was once again clear that the French did not want a rival for the railways passing through their own sphere of influence. Again, by manipulating the loan that they would extend to the Ottoman Empire, they prevented the construction of the Afula–Jerusalem line. The supportive attitudes of the Germans for the Hejaz railway throughout the entire process are covered in Chapters Four and Five.

The last chapter tries to determine the extent to which the Hejaz railway satisfied the expectations placed upon it. Could Abdülhamid increase his reputation in the Islamic world by facilitating the pilgrimage and thus prevent the virus of nationalism from infecting the Islamic elements of the empire? Could the Hejaz railway really enable rapid soldier transfer to the region, thus strengthening Ottoman authority? Did it prove as useful as expected in military terms? To what extent did the Hejaz railway contribute to the economic life of the region?

CHAPTER 1

THE HISTORICAL DEVELOPMENT OF RAILWAY CONSTRUCTION IN THE OTTOMAN EMPIRE

A) RAILWAY CONSTRUCTION IN ANATOLIA AND ROUMELIA

1- *Internal and External Dynamics of Ottoman Railways and the British Lines*

The economic structure of the Ottoman Empire was not mature enough to call for railway construction. Nevertheless, as early as the 1830s, a British officer named Chesney had formulated the very first railway project to be carried out in the Ottoman territory. Chesney's project was one of those initiatives put by the British to reach out to India through the Persian Gulf.[1] As a matter of fact, in 1856, the British inaugurated the Alexandria–Cairo railway with this purpose.[2] The concession of this railway was granted to the British by the governor of Egypt, Abbas Pasha. Later, the Sublime Porte had to bow to the diplomatic pressure exerted by Britain and acquiesce in the construction of this line.[3]

The very first railways inaugurated in the Anatolia and Roumelia regions of the empire were constructed, again by the British, to benefit from the agricultural potential of these fertile lands. German and French lines to be constructed afterwards would serve both strategic and commercial purposes. For the competing imperialist countries, the most effective way of gaining spheres of influence in the Ottoman territory was constructing railways.

The above explanations should not be taken to imply that the Ottoman railways were constructed merely because of the pressures and suggestions of European countries. The Ottomans also hailed this rapidly developing new transportation system as a panacea for their road problems. Starting from Mustafa Reşit Pasha, all bureaucrats of the Ottoman Reform Period, referred to as the *Tanzimat*, who believed in the necessity of political integration with Europe, had underlined the significance of railways for the Ottoman Empire in the proposals that they submitted to the Sultan.[4] The Ottoman rulers were interested in railways primarily due to administrative/strategic concerns. They presumed that escalating internal and external unrest could be averted by rapid deployment of soldiers, which would become possible thanks to railways. Secondly, they hoped that railways could help ease the financial bottleneck emerging as a consequence of the lack of industrialization and that tithe[5] could be increased in tandem with increasing production once the transportation problem was resolved. In fact, in the years 1889–91, while agricultural production throughout the empire increased by 63 per cent, this increase reached up to 114 per cent in regions that were connected to a railway line.[6] That is to say, Ottoman railways can be said to take root from a blend of internal and external dynamics.

The Ottoman Empire not only lacked the capital accumulation to finance railway investments; it was also short of experienced technical personnel. A natural outcome of this fact was that—with the exception of the Haydarpaşa–İzmit line and the Hejaz line— all of the Ottoman railways were constructed by foreign companies. Moreover, the Ottomans even had to offer guarantees to foreign investors, since the trade potential of the lines to be built was not enough in the short run to attract foreign capital.

The first concession was granted to the British in the year 1856 for the İzmir–Aydın Railway. The raw material resources of the region and its potential as an assured market for finished goods provided the incentives for British capital to construct this line. İzmir (Smyrna) was one of the Ottoman Empire's most important ports. It was the only exit point of an extremely fertile hinterland and was a very favorable location for foreign trade.

Even before the commencement of railway constructions, there were already 1,601 British merchants who were engaged in import and export activities in İzmir.[7] This can be explained by the fact that the city had increasingly become an attractive commercial hub for British merchants. Britain had dismantled the Ottoman Empire's last barriers to foreign trade with the Commercial Convention of *Balta Limanı* in 1838. Bans on export, monopolistic practices and internal tariffs were removed,[8] thus eliminating all legal obstacles to entering the Ottoman markets and importing raw material. However, the British merchants faced one major roadblock: transportation. Camel transportation not only increased costs but also brought about a series of other difficulties.[9] Under the circumstances, this problem could only be solved by constructing railways. Robert Wilkin and four of his friends, all British merchants based in İzmir, applied to the Sublime Port to request the concession on a railway to be constructed between İzmir and Aydın. Receiving the backing of the British Embassy, their application was accepted. Since they would not be able to undertake an investment of such scale, they sold their concession to a corporation in Britain. Subsequently, in 1857, the İzmir–Aydın Railway Company was incorporated.[10]

In accordance with the İzmir–Aydın railway agreement, the Ottoman government had to offer a 6 per cent profit guarantee annually against 30.6 million francs of capital invested in the company.[11] Furthermore, the company also enjoyed the right to exploit the land, the mines and the forests that belonged to the government free of charge. In return for these, the first 70 km of the railway line would have to be completed in four years and the Ottoman Empire would not authorize any other company to construct a railway on the specified route.[12]

However, the company could not complete the first 70 km of the railway in the allotted time. In such cases, the agreement entitled the government to seize the 39,600 pounds kept as security deposit and take over the management of the company. Yet, the government decided to extend the period by three years. The reason behind the Ottoman government's granting additional privileges to the company, let alone extending it the aforementioned rights, was the increasing importance attached by the Ottoman rulers to railway construction. The two general reasons why the Ottoman rulers supported railway construction throughout the country were also true for the İzmir–Aydın region. The Western Anatolian region had also been affected by the existing chaos: in addition to conflicts among the Greek, Jewish and Armenian communities, Zeibek, Yuruk and Circassian mobs necessitated rapid soldier deployment. This unrest had an adverse impact on the economic life of the region.

The Ottoman rulers hoped that they could boost production by overcoming the transportation problem and thus increase the yield from tithe. Moreover, railways would improve trade, which would then lead to a rise both in agricultural taxes and customs duties. The construction of the İzmir–Aydın railway took ten years and the line was inaugurated in 1866. As had been predicted, agricultural taxes collected from the regions transected by the İzmir–Aydın railway increased 13- fold between 1856 and 1909. Likewise, the customs revenues collected from İzmir amounted to around 230,000 pounds a year between 1873 and 1877.[13] The İzmir–Aydın line, which was initially 130 km, was extended to Dinar with a supplementary agreement, dated 1879. The length of the line reached 515 km, taken together with its various branch lines.[14]

Meanwhile, the British embarked on the construction of a second railway line that would connect İzmir to Kasaba. When in 1863 a British entrepreneur named Price gained the concession of that line with a 6 per cent profit guarantee from the Ottoman Empire, the İzmir–Kasaba Railway Company was established. The company started its operations in 1864 and in three years completed the 92-km İzmir–Kasaba line.[15] In 1888, the İzmir–Kasaba Company was granted the concession to construct and operate a 92-km line between Manisa and Soma, but without guarantees. The Manisa–Soma line was inaugurated in 1890.[16]

Thanks to the İzmir–Aydın and the İzmir–Kasaba railways, the British gained sway over the Western Anatolian region very rapidly.

Railway construction was followed by an upswing of British trade in the region and British investments shifted to other sectors like mining and municipal services. Besides these, the completion of the railway and the introduction of the land law in 1866, which enabled foreigners to own land, encouraged the British to establish capitalist farms.[17] The expectations of the Ottoman rulers were also met to a great extent. With the resolution of the transportation problem, security was ensured and production increased. The trade volume in İzmir exceeded that of İstanbul.[18] The success of the Aydın and Kasaba lines encouraged the Ottoman administrators to extend these lines further and to start the construction of news ones, primarily towards the Persian Gulf.[19]

However, due to the dependency of production on British industry, crises in the British economy had a direct impact on Ottoman producers. The emery stone is explanatory in the sense that it was produced in Western Anatolia by the British, transported through the British railway and exported to Britain. As for the type of minerals that were entirely isolated and excluded from the regional economy, they included manganese, antimonite, chromium and borax. While the British enhanced productivity in Western Anatolia as a consequence of their intervention, they at the same time rendered the region dependent on British industry. Consequently, the Ottoman economy, which had been articulated with the capitalist world economy through external dynamics, became vulnerable to the cyclical depressions of capitalism. Economic activities in Western Anatolia went through consecutive periods of depression and relative wellbeing following Britain with a time-lag of two years.[20]

Cotton is one of the best examples of products that Britain subsidized in the region to meet its own demand for raw material. The Cotton famine engendered by the American Civil War had compelled British industrialists to subsidize new cotton production areas. The Manchester Cotton Purchasing Union decided to subsidize cotton production in Western Anatolia based on a report drawn up by the British Consul to İzmir. It is interesting to note that in addition to distributing seeds free of charge, providing machinery support to the producers and organizing courses to improve quality, the Union also decided to support the İzmir–Aydın railway line. As a matter of fact, the areas where cotton production rose were all connected to the Aydın railway line.

The British constructed railways also in the European part of the empire with the intention of connecting fertile Ottoman hinterlands to the capitalist world economy. One of these was the Constanza–Cernavoda line, which was granted for 99 years and inaugurated in 1860.[21] The concession of the 220-km Varna–Ruse line was granted in 1861, again for 99 years. Among these, the Cernavoda line remained within the Romanian borders in the aftermath of the war with Russia. The Varna–Ruse railway, on the other hand, was inaugurated in 1866 and then turned over to Bulgaria in accordance with the Treaty of Berlin. Sultan Abdulaziz's first experience with railways was during his journey on this line: according to the *Times*, the sultan was fully satisfied.[22] In addition to the economic boom that they created, both of these railways precipitated the penetration of separatist/nationalist ideas into the region.

2- *A Fruitless Effort: State Railways*
The Ottoman railways that we have examined up to now were all short-distance lines, aimed at mobilizing the trade potential of a limited area. From then on, a comprehensive railway project to connect the Ottoman territory to Europe was on the table. Since the period of Ottoman Reforms, the *Tanzimat*, Ottoman rulers had attached great importance to political integration with Europe. The positive atmosphere created in Europe after the Crimean War and the proclamation of the Reform Decree, called the *Islahat Fermani*, fueled their aspiration to construct a railway line that would tie the Ottoman Empire with Europe. Moreover, Abdulaziz's trip to Europe with Murad and Abdülhamid inspired the sultan and the princes about railways.[23]

At some stage, it was proposed that the government construct a railway connecting the empire to Europe, but this idea was abandoned due to financial difficulties and lack of know-how and technical expertise. Instead, European investor circles were invited. After several initiatives, in 1869, a concession agreement was signed with the Austrian banker, Baron Hirsch. According to the agreement, Hirsch was granted the concession of the 2,000-km railway with a guarantee of 22,000 francs per kilometer.[24]

It soon became clear that Hirsch did not have the financial capacity to run such a project and was merely an adventurer. Actually, Hirsch tried to attract small capital owners by issuing lottery bonds in order to secure the required capital. With this purpose, he issued 1,980,000 security bonds at 400 francs par value with an annual interest of 12 francs. The bondholders stood to win prizes as high as 600,000 francs from the lotteries. [25]

As can be predicted, things did not go as planned. For one thing, the French and British governments did not allow the bonds to enter the Paris and London stock markets. In order to overcome the existing problems, Hirsch and the Ottoman governers came together in 1872 and made significant amendments in the terms of the concession agreement. Most importantly, the railway concession was reduced from 2,000 km to 1,279 km. According to the revised agreement, Hirsch would not have to construct the most demanding and expensive sections of the project. Moreover, the Baron would receive 72,727 francs per kilometer of the lines that he would construct, for a total of 93,017,833 francs. This amount was paid in advance out of the bond yields that were issued. Although the Ottoman Empire would in the end have a railway of 1,279 km, these lines would not be connected with the Austrian network. That is to say, the *Tanzimat*-era pashas' dream of connecting with Europe via railways did not materialize. Publications on that period record that Grand Vizier Mahmut Nedim Pasha and many other administrators were bribed to make sure that the revised agreement would favor Baron Hirsch.[26]

The unfortunate experience with Hirsch[27] did not detract from the importance of railways for the Ottoman rulers. In 1871, Sultan Abdulaziz issued a decree encompassing the idea of surrounding the entire Asian territory with a railway network. The main line was to be constructed between İstanbul and Baghdad. The railway would be connected to the Black Sea, the Mediterranean and the Persian Gulf by extension lines. The mishap with Baron Hirsch had shattered the credibility of private entrepreneurs. Therefore, the government decided to construct the railway on its own. Pursuant to a decree issued on August 4, 1871, the government started laying the rails from Haydarpasha to İzmit. Next year, on October 4, the line had reached only as far as Tuzla. It took two years to connect Haydarpasha with İzmit.[28]

Abdulaziz understood the impossibility of realizing such a gigantic project without a plan. In February 1872, he invited German engineer Wilhelm von Pressel to the Directorate General of Asian Ottoman Railways and commissioned him to design a railway project to meet his goals. Pressel had already proved his merit in fulfilling the missions that he undertook in Germany, Switzerland and the Roumelia railways. In 1872 and 1873, Pressel drew up a comprehensive railway project in line with the sultan's demands. 57,310 francs were spent during Pressel's exploratory visit from İzmit to Ankara. [29]

In the years 1872–73, Wilhelm von Pressel formulated a detailed railway project, which was 4,670 km in length. In line with the sultan's requests, Pressel's project commenced from Haydarpasha and reached Basra via Ankara, Sivas, Mosul and Baghdad. The railway would arrive at the Mediterranean and the Black Seas by branch lines.[30] The cost per kilometer was calculated to be 90,000 francs. In those years, the Ottoman economy was foundering because of the unrestrained debts that it had begun to incur during the Crimean War.[31] Although it was self-evident that the Imperial Treasury, which was drawing closer and closer to the inescapable end, could not finance Pressel's project, they had already started laying the lines from Mudanya to Bursa.

As a result, while the 91-km İzmit line was completed in two years, the construction of the lines between Mudanya and Bursa could not be completed until 1874. The empire spent 185,000 lira to connect the railway to Bursa, however, the line still could not be inaugurated for want of money. The aborted construction was completed only years later in 1895, by granting concession to foreign capital. The government was unable to operate the line that was constructed up to İzmit and leased it to a British company for

20 years. Pursuant to article five of the contract, "the lessee would be sole responsible for the production and maintenance of the line"; therefore the government would not contribute to the financing of any of these expenditures.[32] Apparently, with the bankruptcy of the Ottoman Treasury in 1875, Pressel's project had to be shelved, which marked the end of the dream of constructing a railway by government initiative. When a similar project was being prepared by the Minister of Public Works, Hasan Hilmi Pasha, in 1880, it was already obvious that initiatives of such scale could not be realized without collaboration with European investors.[33]

After 1875, European financial circles ceased to apply for concession to the Sublime Porte. It was natural that capital holders were reluctant to undertake new investments on the territory of an empire which was swirling in a spiral of debt. The revival of European investors' interest in the Ottoman railways after 1881 can be explained by the establishment of the Office of Public Debts, known as *Düyun-u Umumiye*.[34] This office was in charge of collecting the revenues allocated for kilometer guarantees and reimbursing them to railway companies. That is to say, the existence of the Public Debts Council removed the risks associated with the Ottoman Empire's insolvency.[35] The elimination of that risk once again attracted foreign capital to undertake railway investments in the Ottoman territory. The Sublime Porte started to receive applications for railway concessions one after another. Among these applications were British entrepreneurs like Cazalet and French entrepreneurs like Collas.[36] However, Deutsche Bank, which could receive the support of the Foreign Ministry, was also among the competitors and Abdülhamid would opt for the Germans in granting concession. The Anatolian and Baghdad railways, which would be constructed by Deutsche Bank, were to become the engine of Ottoman–German relations.

3- *The Anatolian and Baghdad Railways*
In 1888, after the enthronement of Wilhelm II in Germany, Bismarck's cautious, pro-status quo policy was replaced by an expansionist one. The main reason behind this transformation was

shortage of raw material and inadequate market opportunities, the negative impact of which could be felt more and more in the course of the rapid industrialization process. The lands best-suited to colonization had already been shared out in the last quarter of the 19th century. This had forced Germany to come up with alternative expansionist policies. In line with this "policy of peaceful expansionism" formulated once conventional ways of obtaining colonies had been exhausted, the Ottoman Empire—with its almost untouched raw material resources and the potential demand that could be created by its 25 million people—became crucial for Germany. After the integration of Austrian and Romanian economies to the German economic system through trade agreements, Wilhelm II focused his attention on the vast virgin lands of Anatolia and Mesopotamia. At the same time, Abdülhamid and the Ottoman rulers pinned their hopes for survival on Germany. The sultan's remark, "It would be much better for us and for Germany if they expanded their sphere of influence to the Persian Gulf instead of dissipating their power for exploring useless colonies here and there", clearly indicates that he was ready to make things as smooth as possible for the Germans.[37]

The reason behind Abdülhamid's welcoming attitude towards Germany was the fact that his room for maneuver in domestic and foreign politics was extremely restricted. As discussed above, these were the crisis years for the Ottoman Treasury. Furthermore, the Ottoman–Russian War of 1877–78 had demonstrated that the empire could no longer survive on its own. Britain took over the administration of Cyprus in return for modifying the terms of the Treaty of San Stefano. This showed that Britain was no longer concerned about protecting the territorial integrity of the Ottoman Empire; instead, it was trying to capture the strategically important regions of the empire.[38] Actually, the British invasion of Egypt in 1882 heightened Abdülhamid's skepticism of the British.[39] In addition to Britain and Russia, Austria concentrated on gaining lands in the Balkans, while Italy turned its expansionist eyes towards Ottoman Africa.

Seen in this perspective, there was only one European state remaining which had not yet laid any claim to Ottoman territory:

Germany. The fact that the German Empire did not dominate any Muslim colony was also important for the Ottomans. The European countries'—and primarily Britain's—attempts to provoke Muslim communities to secede from the empire, thus torpedoing Abdülhamid's endeavor to hold at least the Muslim population of his country together, fueled the Ottoman sultan's animosity against these countries. Germans, on the other hand, abstained from such destructive acts against the Ottoman Empire, thereby increasing their popularity in the eyes of the Ottoman rulers. The international climate, so to speak, paved the way for an inevitable rapprochement between the Ottoman and German Empires.

The bilateral relations which had hitherto been only military in nature[40] rapidly assumed economic and political dimensions on the basis of the railways the Germans started to construct in Anatolia during the reign of Wilhelm II. The Germans' railway adventure in the Ottoman Empire started when they were granted the concession of the Anatolian railway. This concession agreement set the conditions for the construction and operation of the İzmit–Ankara line and the transfer of the Haydarpasha–İzmit railway to Deutsche Bank. The tithe revenues of the cities of İzmit, Ertugrul, Kutahya and Ankara, through which the lines would pass, were designated as compensation for the kilometer guarantees, which totaled 15,000 francs.[41] In order to establish a solid relationship with the Office of Public Debts, which would act as the mediator, Caillard, the British representative in the Council, was invited to join the executive board of the railway company.[42]

In addition to Deutsche Bank, Deutsche Vereinsbank and Württembergische Vereinsbank were also members of the consortium. With the participation of Wiener Vereinsbank in 1889, the consortium got hold of the Ottoman railways in the European territory (the İstanbul–Edirne–Plovdiv–Bellova line) by purchasing the majority of the stocks. German entrepreneurs became more enthusiastic about moving forward in the Asian territory after securing the Europe–İstanbul connection. The consortium also gained the concession of the 219-km Salonika–Monastir railway.[43] As a result of these developments, Germans, who had not had a single

kilometer of railway in the Ottoman Empire until 1888, acquired the
concession of a 2,000-km railway network by 1890. The visit by the
German Emperor Wilhelm II and his wife to Abdülhamid in
İstanbul in November 1889 demonstrated that the Germans' designs
on the Ottoman Empire were not limited to railway construction.
This visit can be seen as a commitment to extend political support in
exchange for economic penetration.[44]

In 1890, the İzmit–Adapazari line was completed. At the
inauguration ceremony, in which the managers of German companies
like Siemens and Kaulla participated, Minister of Public Works Raif
Pasha expressed his wish that the Germans extend the line to
Baghdad. Ambassador Radowitz, in his report to Prime Minister
Caprivi, stressed Abdülhamid's willingness to initiate negotiations
with the managers of the Anatolian Railway Company in order to
extend the line to Mesopotamia. When Kaiser Wilhelm read the
report himself, he specifically expressed his contentment.[45]

The first train of the Anatolian Railway Company reached the
Ankara station within the timeline stated in the contract, namely on
November 27, 1892. Deutsche Bank was again the winner of the
concession competition against British and French entrepreneurs in
the continuation of the Anatolian railway. The British, the French
and the Russians reacted very harshly to the granting of the
concession again to the Germans. The French on the one hand tried to
gain the concession for the extension of the Anatolian railway
through the support of their Ambassador Chambon and on the other
to prevent Deutsche Bank from gaining the concession by making
intrigues in Yildiz Palace.[46] British Ambassador Clare Ford officially
expressed his opinion to the Sublime Porte that the granting of the
concession for Konya to the Germans would impair the interests of
the British and threatened Ottoman rule by organizing a
demonstration with the British fleet on the İzmir coast.[47]

In accordance with the agreement, the lines would be extended
from Ankara to Eskisehir–Konya and then Kayseri. The Kayseri line
was never built. 13,800 francs, the kilometer guarantee calculated for
the Eskisehir–Konya railway line, were reimbursed from the tithe
revenues of Trabzon and Gümüshane. The Office of Public Debts

continued to serve as a go-between for the collection of taxes and their submission to the companies in the form of kilometer guarantees. The Eskisehir–Konya line was successfully completed in 1896, as stipulated in the contract. Henceforth, it would be possible to reach Konya from İstanbul in two days.[48]

It was already known that after the extension of the Anatolian railway to Konya, the issue of the Baghdad railway concession would be on the negotiating table and that the competition would be keen. If Germany won the Baghdad railway concession, this would mean the expansion of its influence to the Persian Gulf. Britain, France, and Russia felt uneasy at this prospect: the British, because of the passage to India; the French because of the area of influence that they were trying to establish in Syria; and the Russians because of their intention to expand their influence to the South. Thus, all were trying their level best to make sure that the Baghdad railway concession was not granted to the Germans.

In 1897, Wilhelm II demonstrated the significance of the Ottoman Empire in Germany's new foreign policy by assigning one of his most trusted statesmen in the formation and implementation of *Weltpolitik*[49], Marschall von Bieberstein, as Ambassador to İstanbul. Marschall worked very hard to strengthen Germany's sway over the Ottoman Empire during his 15-year tenure.[50] From what Marschall wrote in his 46-page report to Chancellor Hohenlohe, it was obvious that he attached great importance to the Baghdad railway.[51]

> ...If Germany continues to increase its economic influence in the East, if the Haydarpasha Port is built in order to expedite the transportation of German goods by German vessels and the Anatolian railway is extended to Baghdad as a German initiative using exclusively German materials, the claim that "the whole of the Balkans is not worth the bones of a single Pomeranian grenadier" will become a moment in history, which has lost its contemporary validity.

Marschall was referring to the famous quotation from Bismarck, trying to underline that the cautious policy of the past was now obsolete. In 1898 Emperor Wilhelm's second visit to the Ottoman Empire turned into a real display of power. His renowned Damascus speech, in which he declared himself the guardian of 300 million Muslims, created quite a stir in Europe. When a British journalist making observations during the journey heard these words, he could not help but say, "Hopefully this friendship will would not extend all the way down to India."[52]

The immediate outcome of Wilhelm II's visit was the granting of the Haydarpasha Port concession to the Anatolian Railway Company, which had been trying to secure it for a long time.[53] A few months after gaining this concession, the company applied to the Sublime Porte for the Baghdad railway. The British filed a counter-application through a Hungarian banker named Rechnitzer. Rechnitzer's group, receiving the support of Damat Mahmut Pasha and the British Ambassador Sir O'Conor, did not demand a kilometer guarantee in their proposal.[54] This was quite astonishing, for it was impossible to accomplish such a gigantic project without any kilometer guarantee. Moreover, it was already clear that Abdülhamid would not grant the concession to the British. So apparently Britain's ultimate goal was not to win the concession but rather to drag out the application procedure for the Germans.

Marschall, seeing that their chances of winning the concession were getting slimmer, employed an astute tactic and ensured that a preliminary agreement was signed between Deutsche Bank and the Ottoman government. The main purpose of this interim agreement was to bypass the competing powers. It should also be mentioned that Deutsche Bank had lent the Ottoman government 3,500,000 marcs at an interest rate of 7 per cent prior to this preliminary agreement. The preliminary agreement, dated December 23, 1899, did not cover details like the kilometer guarantee, but only stated that the Baghdad railway concession was granted to Deutsche Bank.[55] At the end of this process and the final agreement, concluded on March 5, 1903, Deutsche Bank was given the concession for the construction and operation of the Baghdad railway for 99 years.[56]

The new concession agreement provided very important privileges to the newly-established Baghdad Railway Company, including the right to run maritime operations in the Euphrates and Tigris rivers.[57] Since 1831 this had been the exclusive right of the Lynch Company, a British firm. Furthermore, with the new agreement, the Germans obtained an additional privilege on the ports of Baghdad and Basra. These articles greatly augmented British opposition to the Baghdad railway. As has previously been mentioned, controlling the Persian Gulf was the first and foremost concern for the British, in order to ensure the security of the road to India.

The Baghdad Railway Company was authorized to utilize the water resources and conduct mineral exploration activities within 20 km on both sides of the lines that they would lay.[58] Unlike the İzmir railways, the aim with the Baghdad railway was to exploit the abundant untapped potential of the region. This entailed the establishment of infrastructural facilities along the railway. German entrepreneurs were aware of this fact. Even before the conclusion of the agreement, a visitor named Hermann had been commissioned to investigate the agricultural capacity of the region through which the line would pass. The company management had also stepped up geological explorations in the region. It was predicted that the area between Adana and Aleppo was rich in iron and tin, Diyarbakir had plenty of silver and Kirkuk had important oil reserves. The reports prepared by researchers substantiated that the Kirkuk region had comparable potential to Baku in terms of oil reserves and that, thanks to the Baghdad railway, these energy resources could be utilized efficiently. After the collection of these data, Deutsche Bank gained the privilege of exploring and drilling for oil in the region.[59]

In the aftermath of the 1903 agreement, the first 200 km—from Konya to Bulgurlu— were completed in 19 months, since the area was ideal for railway construction. However, this promising first move could not be sustained and for five years no further progress was made.[60] The lines immediately beyond Bulgurlu were the most costly section. A capital of around 600 million francs was required for the completion of the Baghdad railway and even Deutsche Bank administrators admitted the impossibility of procuring this sum from the German monetary market. On the other hand, the interest charges and redemption costs of the Baghdad railway bonds issued for the Konya–Bulgurlu line had put a great strain on the Ottoman Treasury. The empire could no longer incur new debts in order to pay the total 31 million franc annual kilometer guarantee.[61] Deutsche Bank's desperate efforts to incorporate British and French capital as partners in the Baghdad railway project proved vain. Nevertheless, it should be noted that Deutsche Bank did not sit idle even during the years that the construction was suspended. For instance, in 1907, the bank gained the concession to irrigate the Konya Plain and as a result of the irrigation and land improvement activities, the Germans' cotton yield in Anatolia outstripped that of all of its colonies in Africa.[62]

On June 3, 1908, an additional bilateral agreement to resume the construction was signed. The agreement set the terms for the construction of an 840-km line between Bulgurlu and Al Halif, which was the most expensive and challenging section because of the Taurus and Amanos mountains. The loan extended by the Baghdad Railway Company would resolve the issue of kilometer guarantee.[63]

However, with the Young Turks' coming to power in 1908, the Baghdad Railway Company started to face serious difficulties. Baghdad railway workers, enjoying the relative freedom brought about by the constitutional monarchy, went on strike on September 14, 1908.[64] The Young Turks had even started to question the validity of the agreements between Abdülhamid's government and the Germans. In 1909, upon the proposal of the Baghdad deputy İsmail Hakki, the Baghdad railway project was debated in Parliament. Dissident deputies were lambasting the railway concessions. There were various grounds for criticism, and some of them centered on the financial dimension of the concession. Krikor Zohrap, İsmail Hakki and Cavit Beg made a point of demonstrating that the concession was very lucrative for the railway company due to the amount of money that the Ottoman Empire had to borrow. The deputies used the Konya–Bulgurlu line as an example. The financing of this railway was created by selling 54 million francs' worth of bonds. The interest payment on the bonds was reimbursed from the tithe of Konya, Aleppo and Urfa provinces.[65] That is to say, the Baghdad Railway Company did not have to pay a penny. The activity of the company was limited to the issuance of bonds. With the

Baghdad railway case, dissident deputies were able to observe that foreign railway companies operating in the Ottoman Empire had overcome the problem of financing by borrowing from the bonds issued by the Ottoman government.

The anti-German wind blowing in the Ottoman Empire after 1908 subsided rapidly. One of the main reasons behind this was the tactless attitude of the British and French in trying to transform the changing conditions into a means of exploitation. When in 1910 the Minister of Finance, Cavit Beg, turned his eyes to Britain and France in order to borrow money the Treasury was in dire need of, the two powers set humiliating conditions which would render the Ottoman Empire fully dependent on them.[66] The Germans did not miss the opportunity and immediately extended a helping hand to Cavit Beg. In the aftermath of the loan agreement with Deutsche Bank, Cavit Beg reported that the Germans had run the entire process very discreetly and had not stipulated any condition that would dishonor Turkey.[67] Evaluating the outcome of the loan agreement, Helfferich said, "The status we achieved during the constitutional regime was even far beyond the [privileges] we had enjoyed during Abdülhamid's tenure."[68]

Following the loan agreement with Deutsche Bank, Baghdad railway negotiations resumed. The Germans seemed willing to arrive at a compromise concerning the objections raised against the former agreement. The negotiations were finalized with a new deal signed in 1911.[69] Undersecretary Kiderlen hailed this agreement as "the great victory of German entrepreneurial spirit despite all the obstacles set by the great powers and the unexpected changes that the Turkish administration went through".[70] As a result of the agreement, the company immediately embarked on construction activities in a number of locations simultaneously, including Bulgurlu and Baghdad. By the start of World War I, 887 km of railway starting from Konya had been completed. All that remained was a 38-km section in the Taurus mountains, a 100-km section in the Amanos mountains, and the 690-km Samarra–Mosul–Tel Abyad line.[71]

Considerable attention has been paid here to the Anatolian and Baghdad railways because these lines were not only critical in military, political and economic terms, but also because they were complementary to the Hejaz railway, which is the actual topic of this study. As a matter of fact, Abdülhamid had announced the Hejaz railway project in 1900, when negotiations over the final concession agreement for the Baghdad railway were still underway. Later in this study it will be demonstrated that for Abdülhamid the two railways represented a single whole. The Syrian railways, the subject of the next section, is also intimately connected with the Hejaz railway. Syria is the starting point of the Hejaz railway. Moreover, during the construction of the Hejaz railway, significant problems were posed by the involvement of French railway companies and French diplomatic agents in Syria. In order to better understand these problems, let us first take a closer look at the railways that the French constructed in Syria.

B) IMPERIALIST DILEMMAS AND SYRIAN RAILWAYS

1- *The Origins of French Influence in Syria and the French Railways*

Syria was very heterogeneous in terms of ethnicities and religions. The ethnic composition of the region was a blend of Arabs, Assyrians and Phoenicians, which later on amalgamated with Greeks, Romans, Crusaders and Turks. At the end of the 19th century, Syria was one of the regions of the Ottoman Empire where fierce struggles for influence raged. The German march towards the Persian Gulf through Anatolia by means of the Baghdad railway increased the prominence of Syria for both the British and French. The French regarded the region as their historical area of influence. However, the British did not want to submit to French domination in the region since Syria was the East Mediterranean gateway to the Road to India. In their struggle for influence, both the British and the French resorted to the same method, which they frequently applied during the second half of the century: they sought to obtain railway concessions while at the same time acting as the guardian of one of the communities in the region, with the aim of securing their commitment and loyalty. To this end, the French became the protectors of the Maronites, while the British tried to win the Druze to their side.

The seeds of the relationship between France and the Maronites were sown during the Crusades. After François I had gained privileges from Suleyman the Magnificent in the form of capitulations, French kings started to enjoy greater religious sway over the Christian subjects living in the Ottoman Empire. The very first Catholic missionaries in the Ottoman Empire were French priests, who were allowed to perform their missions unhindered in Syria and Lebanon. Those missionaries established schools in the region and tried to disseminate French Catholic culture. At the same time, French merchants established chambers of commerce in Aleppo, Alexandretta, Tripoli and Beirut and acted as intermediaries for Europe in trade with Syria.[72]

In the aftermath of the Syrian Crisis of 1860–61, Beirut became the most important port and a leading commercial hub in Syria. The French, who were strengthening their political and economic presence in Beirut and in the region, had contributed significantly to this rapid progress. The French had turned the Syrian Crisis of 1860, with its international dimensions, to their best advantage and increased their influence tremendously.

The tensions that emerged between the Druze and the Maronites in 1860 resulted in Druze plundering and murdering the Christian population in the Lebanese mountains and Damascus. When news of the incidents reached Europe, Napoleon III mobilized a 5,000-strong military force in the region.[73] According to diplomatic circles, the French troops were deployed to help the sultan combat the rebels. As a matter of fact, Napoleon believed that it was high time to render French dominance in Syria indisputable and perhaps to realize the project of an independent Syria. Moreover, France was having a hard time supplying raw material for its knitting industry. Particularly silk knitting, one of the engines of the French economy, was having difficulties in finding raw silk. French knitters were exerting pressure on the government to boost silk production in

Syria and underpin French investments in the region.[74] In the 1870s, French industrialists were employing around 1,200–1,250 people in silk knitting in Syria.[75] Beirut was the hub of silk and particularly of the raw silk trade. Most of the raw silk produced in Syria was sent to France.[76]

Not surprisingly, however, Britain was intent on curbing the further expansion of French influence in Syria. Realizing that they would not be able to interfere with the advance of French soldiers towards Beirut under the pretext of protecting the Maronites, the British tried at least to restrict the number of soldiers and prevent them from marching into the Syrian hinterlands. When the Maronites, heartened by the advance of French troops in Beirut, attempted to retaliate, the British again seized the opportunity to safeguard the Druze. Meanwhile, as the conflict spread to Damascus, the resident Christians immigrated to Beirut.

Minister of Foreign Affairs Fuat Pasha, who was dispatched to the region, claimed that the French were involved in forcing the Christians to emigrate by marking their houses. According to the Pasha, the French were intentionally fomenting the emigration, which would then serve as a pretext for the French troops' continued presence in Beirut.[77] However, Fuat Pasha had already punished the perpetrators and those who were found guilty of negligence and taken protective measures for Christians. Furthermore, he levied a special tax on the Druze, who had attacked the Maronites, to compensate for their loss. Yet, Napoleon III, disregarding the fact that the incidents had been contained, dispatched soldiers to Beirut on August 16, 1860, under the pretext of helping Fuat Pasha and relying on the historical bonds between France and the Maronites. Consequently, Lebanon began to be governed as a "privileged administration".[78]

As a result of the incidents of 1860–61, the French were firmly established in Beirut and started to obtain concessions to construct and operate transportation lines towards central Syria. For instance, in 1863, a French company constructed the Beirut–Damascus highway and turned a significant profit. In 1893–94, after the new port had begun to operate, commercial activity in Beirut soared. Meanwhile the French were extending intensive support to boost the development of the city in all fields. The level of economic and social development revealed itself in the rising population: 40,000 in 1860, 100,000 in 1885 and 160,000 in 1914.[79] The most important administrative/economic center of Syria, Damascus, was gradually being dethroned by Beirut.

In the midst of the 1880s, various railway projects were being drawn up to link Beirut with the agriculturally fertile hinterland of Syria, thus enabling the transfer of goods to the Port of Beirut. With this purpose, in 1891, Hasan Beyhum Effendi, a prominent figure in Beirut, gained the concession to operate a steam tramway between Beirut and Damascus. The concession, which was granted for 99 years, was immediately transferred to the newly established French company, "Compagnie de la voie ferrée économique de Beyrouth Damas". Moreover, another company, called "Compagnie des tramvays de Damas et voies ferrées économiques de Syrie", had already gained the concession to construct and operate the Damascus–Muzeirib line three years before. Muzeirib was the center of grain production and the heart of the caravan routes in Syria. The two aforementioned companies merged under the name of "Société des Chemins de fer Ottomans économiques de Beyrouth-Damas-Hauran en Syrie".[80] In 1893, this newly merged company was able to gain the concession for the Damascus–Houmus–Hama line, for which there had long been intense rivalry. While there was no kilometer guarantee for the arterial railway, a guarantee of 12,500 francs was given for the Damascus–Birecik extension.[81] By virtue of the new concession, the French would be able to extend the Beirut–Damascus railway from Damascus towards Anatolia in the north and Arabia in the south. It should be mentioned that the company in question gained new concessions and assumed the name "Société Ottomane du Chemin de fer de Damas-Hamah et Prolongements", or "DHP" for short, in 1900.[82] Later, the company managed to increase the kilometer guarantee for the Rayak–Houmus–Hama line to 15,000 francs.[83] In return, they would have to connect this line with the Baghdad railway should the Ottoman administration so desire.

Construction work on the 100-km Damascus–Muzeirib line began in late 1891. With this line, the Hauran region, which was the breadbasket of Syria, would be linked with Damascus, a commercial center. Furthermore, export activities that used to be carried out directly through Acre and Haifa could probably then be carried out via Damascus through Beirut. The French managed to complete the Damascus–Muzeirib section of the line by 1894. The travel time from Damascus to Muzeirib, which had previously been two and a half days, was reduced to three hours after the inauguration of the line.[84]

The following year, the Beirut-Damascus section—a laborious one—was also completed, connecting the Damascus–Muzeirib line with the Port of Beirut. Surmounting the Lebanon and Anti-Lebanon Mountains was grueling and costly. Passenger trains could only travel at a speed of 18.5 km per hour, and freight trains only 10 km per hour.[85] As a consequence, the distance between Beirut and Damascus, which was 147 km, took 12 to 13 hours. The other handicap that the French had to tackle was the fact that they had to construct the Rayak–Aleppo section, for which they gained concession, in the form of standard gauge, while the lines in the Beirut–Damascus–Muzeirib section were narrow-gauge. It was obvious that this would create a big problem in terms of freight and passanger transportation.[86] This also constituted a setback for the troops to be dispatched to Syria. [87]

The very first railway that the French inaugurated in Syria was the 87-km Jaffa–Jerusalem line. It was Yusuf Navon, an Ottoman subject, who acquired the concession for the Jaffa–Jerusalem line in 1888 for 71 years.[88] It should be noted that the actual owner of the Jaffa–Jerusalem railway concession was a German banker named J. Frutiger, who resided in Jerusalem. Frutiger must have preferred to stay in the background for legal and tactical reasons.[89] In 1889, the concession for the Jaffa–Jerusalem railway was transferred to the joint venture company, "Société du Chemin de fer Ottoman de Jaffa à Jerusalem et Prolongements", which was established in Paris by an entrepreneur named Collas. Almost all of the shares of the company, which were issued in the Ottoman language, were held by French people. The Jaffa–Jerusalem railway started to operate on September 26, 1892. The railway company's main source of income was

Christian pilgrims and tourists visiting the Holy Land. Since freight transportation between Jaffa and Jerusalem was insufficient and the pilgrim and tourist traffic only seasonal, the company could not achieve its financial targets and started to encounter funding problems. For this reason, the Jaffa–Jerusalem Railway Company had to temporarily suspend its operations in 1894[90] but thanks to a restructuring process, it was able to survive and make a profit in the following years.[91]

Finally, it should be mentioned that in 1890, another French company, "Société Ottomane des Tramvays Libanais Nord et Sud de Beyrouth", acquired the concession for the construction and operation of a steam tramway between Tripoli and Sayda. The company had taken over this concession from a Syrian landlord called Hadra. The tramway company was able to inaugurate a 19-km tramline, which could operate only locally, in 1898.[92] For lack of financial resources, this line, which operated between Beirut and Maamelteyn,[93] could not be extended northwards.[94]

2- *The Haifa-Damascus Railway and Other British Projects*

The most important British initiative in Syria is the Haifa/Acre–Damascus Railway project. Haifa, the natural harbor of the region, constituted the ideal starting point for a railway that would lead from the coast to Damascus. The lines would pass through the most fertile—and best-suited to railway construction—lands in Syria. Analysts of the time envisaged an upswing in Syrian exports and agricultural production once a rail connection was established between the coast and the region that covered the Hauran and Jordanian plains.

In 1882, the Sursock brothers acquired the concession to construct and operate a railway between Acre and Jordan. However, since they could not find the required capital, their concession became null and void.[95] In 1890 the concession for the construction and operation of the 237-km Acre–Damascus line was "granted to Monsieur John Robert Philling and Yusuf İlyas Effendi, an Ottoman subject".[96]

The concession granted to the British for the Haifa–Damascus railway in 1890 raised a number of concerns for the French. For example, in 1893 the French consul to Haifa, Bertrand, stated that due to the concession held by the British, the DHP company would have to brace for fierce competition. As a matter of fact, the Haifa–Damascus line had significant operational advantages over the Beirut–Damascus railway. The distance being shorter and the land much more suitable for railway construction, travel time and passenger and freight tariffs would be lower.[97] The reports drawn up by the French consul to Beirut echoed similar concerns.

These anxieties led the French diplomats to warn the Ottoman authorities against the activities of the British in Syria. They tried to have the privilege canceled by stating that its owner, Yusuf İlyas Efendi, was actually nothing but a pawn in the process, that it was the British government who stood to profit, and that the secret aim of the British was to take hold of Syria, following Cyprus and Egypt.[98] It was reported to the sultan that the British desired to gain military power in Syria, using Yusuf İlyas as a tool.[99]

Chief Secretary Süreyya Pasha evaluated the arguments put forth by the French dispassionately. In discussing the subject with Abdülhamid, he emphasized that Yusuf İlyas and other local authorities were known to be in the habit of acquiring such privileges and selling them to foreigners with the aim of making a profit. On the other hand, he said he did not find realistic the claim that the capital needed for the railway would be provided by the British government. That argument, he held, derived from French anxiety about a potential loss of influence in Syria and meddling in their affairs on the part of the British. According to the Pasha, competition was to be encouraged so that Syria did not end up being ruled by a sole power. Privileges should be granted primarily to countries such as America, Germany and Belgium, who, the Pasha believed, were not striving for any political power in the region.[100] In the end, as had been expected all along, the French could not prevent the sultan's awarding the concession to Philling via Yusuf.

It has since become clear that the British authorities could not accept the continuation of French power in Syria. For instance, one of the chief figures behind the Sykes/Picot Agreement, Mark Sykes, speaking in the House of Commons, clearly expressed his reaction against the French: "These (railway) concessions, which have been extracted from Turkey ... mean a monopoly of all Syrian transit." According to Sykes, such concessions "must, whether the financiers desire it or not, pave the way to annexation".[101] Sykes' views are supported by a speech made in Parliament by Mr Lloyd:

> France, of course, has had for years past certain economic interests to consider in the form of railways, and for us to have embarked on a policy of trying to convert Syria into a British sphere of interest would no doubt have been an unfriendly policy towards existing French economic interests.[102]

The railway, which was to be constructed by the English company to standard gauge, would extend from Acre to the neighbouring harbor of Haifa, and the concession would be valid for 99 years. Construction was to commence within 15 months of the conclusion of the concession agreement and be completed within a year. Unless these conditions were fulfilled, the concession would become invalid.[103]

From the very beginning, the British entrepreneur Philling encountered difficulties in the construction of the Acre–Damascus railway; all his initiatives ended in fiasco. Philling's concession was nullified several times. The ground for the nullification was the fact that the construction did not begin on time. Nevertheless, the British entrepreneur was able to secure a new concession in 1891. The Syria Ottoman Railway Company which was incorporated at the end of that year managed to construct a railway line of merely 8 km and had to suspend construction in 1893 due to a financial bottleneck.[104] Meanwhile, Philling's efforts to gain a kilometer guarantee from the Ottoman government bore no fruit. In 1895 Philling was able to secure a new extension on the condition of completing the line in the course of three years;[105] however no progress was achieved during this period.

In 1898 Philling was able to obtain another extension from the
Minister of Public Works, Mahmut Celalettin Pasha. In the
following year, however, Mahmut Celalettin Pasha died and his
successor Zihni Pasha declared the extended concession null and void
on the grounds that it had not been approved by the sultan.[106]
Meanwhile the railway construction was progressing very slowly.
During the summer months, nine Italian railway workers died of
malaria and several others were admitted to hospitals in Haifa and
Nazareth, which held up the construction activities again.[107]
Consequently in 1900, despite tremendous efforts expended by the
British Embassy in İstanbul, termination of the railway construction
could no longer be averted.[108]

Philling's failures in constructing the railway undoubtedly
contributed to the Ottoman administration's final decision. Political
factors also affected the decision, including a general lack of confidence
in the British on the part of Abdülhamid and the Ottoman
administrators, and the undisputed French influence prevailing in the
region. Another, equally crucial factor, was the fact that Abdülhamid
had decided to embark on the construction of the Hejaz railway. The
Ottomans wanted to establish their own control over the
Mediterranean exit of the Hejaz railway, which would be constructed
by government initiative. As we shall see later, a connection line with
the Hejaz railway was constructed between Haifa and Dera.

In the aftermath of Philling's fruitless initiatives, the British
formulated yet another railway project. Black and Fraser proposed a
gigantic railway project which would start from Ismailia in Egypt
and extend all the way down to the Persian Gulf via Aqaba.[109]
Considering that it was the shortest way to link the Mediterranean
with India, the politico-strategic importance of this line for Britain
was self-evident. But as to its feasibility, even the British experts
expressed concerns, raising the following questions: first of all, how
could such a long line be financed? Given that the lines would pass
through regions whose trade potential was estimated to remain poor
even in the long run, what were the odds of the company's making a
profit? How could security be established in the deserts of Arabia,

which lacked political stability and were inhabited by Bedouin tribes, etc.?[110]

Moreover, it was also obvious that the Ottoman administration would not welcome this project. It was impossible for the sultan to accept a project which would maximize British influence in the region, since it was known that Britain intended to establish the security of the road to India by gaining control over certain strategic locations. Taking into account the significance that Abdülhamid attached to the office of the caliph, fear that British influence would expand in the Hejaz hampered the project. All these factors explain why the Railway to India Pioneer Company, Ltd., founded in London to undertake the project, was stillborn.

Although it never proved possible to connect Egypt and the Persian Gulf by railway, this project never left the British agenda. Richard Hennig successfully analyzed the reasons behind the Egypt–Persian Gulf projects in 1912. It is useful to quote here some of the highlights of his article, for they represent the interesting observations of a German writer concerning the issue. Hennig mentions the recent increase in the number of railway projects in Britain that would connect the Red Sea with the Persian Gulf. Financial aspirations, he claims, fail to explain such a glut of railway projects passing through the deserts of Arabia. How could one justify the plausibility of those projects, which called for investments as high as 170 million francs? Hennig goes on to answer his own question: of course the reasons are politico-strategic. Here is how Hennig continues his analysis:[111]

The British aspiration was to connect Africa and India, which were under their rule, and integrate them. After consolidating its position in Kuwait, turning Southern Iran into its stronghold, strengthening its influence in Mesopotamia and conquering Egypt, Britain was disgruntled about the fact that Arabia stood like a wall separating its colonial regions in North Africa and South Asia. Moreover, the Ottoman influence was weakening in Arabia, such that some of the tribes living in the inner parts of Arabia were not even aware that they were subject to a sultan on the Bosphorous and the rest were hostile towards the Ottomans. In short, the British were now aiming to terminate the already-weakened Ottoman domination in Arabia. For instance, it was not possible to interpret the incessant insurgencies in Yemen without reference to British backing.

Hennig was questioning why Arabia, one of the least significant countries in the world economy, was so vital for the Ottoman Empire and Britain. For the sultan, as the spiritual leader of Muslims, the prospect of parting with the holy cities was truly distressing. For Britain, holding sway over the Holy Land would mean burnishing its reputation in Islamic countries, which constituted a majority of its colonies. British railway projects across the desert could serve this end. Furthermore, they would be able to establish military control over a vast region from Khartoum to Calcutta and deploy their troops from Suez to Kuwait in three days. That is why they aspired to build a railway from the Nile all the way to Arabia, Mesopotamia, Iran, Baluchistan, even India and China. The British were also formulating alternative projects that would serve the same purposes. For instance, a railway project which would connect Jeddah, on the Red Sea coast, to Kuwait, on the Persian Gulf, was on the agenda. Nevertheless, the Ottomans would on no account grant concessions for such railway projects to the British. They considered it suicidal.[112] In the face of these facts, unless the British conquered Arabia, the chances of a railway that would connect Alexandria and Kuwait or Jeddah and Kuwait were nil. Hennig's analysis of these desert railway projects devised by the British was clearly realistic. It should also be emphasized that the British never abandoned these desert railway projects. For example, in 1911, Charles E. D. Black from the Geographical Department of the India Office was still proposing a railway project which would commence from Port Said and lead all the way to Basra—even to India—through Aqaba and Maan.[113]

This final section of this chapter summarizes the railway projects originating from the Ottoman Empire which were intended to connect Egypt with Syria. The first creator of such a project was the

Ottoman Commissar responsible for Egypt (Extraordinary Commissar to Egypt), Ahmet Muhtar Pasha. As early as 1893, the pasha submitted a report to the sultan emphasizing the drawbacks of telegraph lines' being operated by the British in regions like the Hejaz and Yemen, and urging him to explore ways of laying an alternative line.[114] A letter of request that he wrote to Abdülhamid in 1897 demonstrates that the pasha continued to come up with radical ideas. He then suggested the simultaneous construction of two railways, one from Konya to Damascus and the other from Damascus to the Suez Canal. According to the Pasha, enabling the Ottoman troops in Syria to pass through the Sahara Desert by railway would undermine British domination in Egypt, which in turn would bolster the military power and political sway of the Ottoman Empire in the region and automatically ameliorate the problem of defending the holy cities. Muhtar Pasha's remarkable insight came in anticipating that the British would try all possible means to forestall any railway project between Damascus and the Aqaba Gulf.[115] The radical reactions of the British—including even a threat of war—when the Ottomans did attempt to construct the Aqaba line, will be discussed later.

After Abdülhamid's rule had ended, the Young Turks and the British agreed on a similar project. However, there were serious concerns as to the railway's competitiveness against the sea roads on the Suez Canal. As a matter of fact, it was calculated that sea transportation took only two hours longer than rail passage, which was too feeble an argument to justify a 300-km train journey under the blazing sun. In 1912 in his comprehensive coverage, Fritz Lorch asserted that, despite the recent economic progress in Jaffa and Gaza, the railway in question could still not yield a profit. In the Gaza–Alexandria section of the Jaffa–Alexandria line, he observed, the only noteworthy residential area was Al Arish. For this and similar reasons, the British railway company shelved the construction project for good.[116]

CHAPTER 2

DECISION TO CONSTRUCT THE HEJAZ RAILWAY

A) DECLARATION OF THE IMPERIAL FIRMAN CONCERNING THE HEJAZ RAILWAY AND INITIAL REACTIONS

Abdülhamid's imperial will (*irade-i seniyye*) which, in the German orientalist Hartmann's words, "astounded the entire world",[1] was issued on May 1, 1900, the sultan's birthday. The project sounded unusually ambitious because it was to be built and operated entirely by the Ottoman Empire. Foreign capital would not be raised; the "sacred line" would be a purely Muslim project, built by Muslim engineers using only local materials.[2] Non-muslims and foreigners were prohibited from disposing of land and property along the line that led from Damascus to Mecca.[3]

Abdülhamid's explanations particularly astounded the Europeans, and were considered most unconvincing. Taking into account the unfavorable economic and political circumstances of the empire on the eve of the 20th century, it was quite understandable that the project was regarded as unrealistic. An article entitled "Die Mekkebahn" published in *Die Reform* magazine, is a good representation of Europeans' approach to the Hejaz railway. The writer expressed his disbelief in the initiative as follows:

A country subjected to international supervision due to its failure to fulfill its obligations towards its creditors, a country which has to earmark a significant portion of its revenues to the payment of outstanding debts and interest expenses; a country which lacks the necessary resources and qualified staff to manage any economic enterprise such as railway, road or port construction, is embarking on such a comprehensive project – which will obviously not make any profit in the foreseeable future![4]

The reports of European diplomats in the Ottoman Empire also stressed the implausibility of the project in a rather cynical way. Even Marschall von Bieberstein, an ardent supporter of the Baghdad railway, stated that "no one in his right mind can have faith in the feasibility of this 1200-kilometer 'sacred line'".[5] In his next report, written three days later, Marschall pointed to Foreign Minister Tevfik Pasha's misgivings about the project. He thought that the Ottomans lacked not only the finances needed for the construction but also the operational capability to run the railway, should it ever be completed.[6] The chief secretary of the consul, Wangenheim, shared the same viewpoint. In his report, Wangenheim rather sarcastically wrote, "Wooden traverses to be brought from Macedonia will later be used by the Bedouins for heating."[7] The reports of British and French consuls based in the region indicate that they were of the same mind as their German counterpart. For instance, one of the French consular reports suggested, "This panislamic project of constructing a railway from Syria to Mecca is a utopia which will never come to fruition."[8] Richard, the British consul to Damascus, expressed his disbelief in this extensive project due to lack of adequate funding. The Austrian ambassador likewise stated that the project was being ridiculed in financial and political circles. According to the ambassador, Abdülhamid's aim with the Hejaz railway project was to calm the reactions to railway concessions given to the Russians.[9]

Some of the Ottoman administrators whom Abdülhamid had consulted also explained that the state was not technically and financially capable of undertaking a railway project. There were those who claimed that, even though the Hejaz railway might, against all odds, be built, the expected benefits would not be reaped and it could even create a security gap. For instance, Grand Vizier Said Pasha believed that an imminent attack from the Red Sea could put the holy cities under the threat of invasion. Moreover, because the deserts were not "areas for settlement" the railway would not bring any economic benefit.[10]

B) THE MAIN FACTORS BEHIND THE DECISION TO CONSTRUCT THE HEJAZ RAILWAY

1- *Economic Factors*

What were the reasons behind Abdülhamid's decision to embark on the Hejaz railway construction against all odds? Naturally economic factors are the first to come to mind in the case of railway projects. Suffice it to say that such a perspective does not help us at all in understanding the motivation behind the Hejaz railway. Indeed, the Hejaz railway could well be the only railway project in which economic factors did not play a role at all.

As a matter of fact, Hejaz province was a financial burden on the shoulders of the Ottoman Empire. The state did not generate any significant revenue from the region other than customs duties collected in Jeddah. This amount was far from covering the administrative and military expenditures of the region. Moreover, half of the customs revenues of Jeddah were allocated to the amirs of Mecca.[11] Apart from the customs revenues, Ottoman sultans use to send 25,000 kurush to the Amir of Mecca every year out of their personal budget. This amount was called the *atiye-i hümayun*. In addition to these, the people of the Haramain (Mecca and Medina) received a considerable amount of money, called *surra*, from İstanbul every year during the hajj season. *Surra* was divided amongst people according to the *surra* ledger. A *surra* regiment would depart from Üsküdar following a lengthy state ceremony.[12] Once in Mecca, *surra* sacks would be distributed to their rightful owners as specified in the *surra* ledger under the oversight of the amir of Mecca, sheik-ul

Haram, trustee of the *surra*, and *qadi* of Mecca.[13] Furthermore, to prevent Bedouins from attacking hajj groups, some Bedouin tribes were supplied with grains and paid a certain amount of money called *urban surra*.[14] The Ottoman Empire followed the tradition of sending this money regularly, even during crisis years for the treasury.

Obviously the holy cities were quite costly for the Ottoman Empire. Because the Hejaz land was sandy and stony, the region was not self-sufficient in terms of agricultural production and was dependent on external supplies. The Austrian consul, Toncic, made a striking analysis of the region in his report, in which he defined the Hejaz as the poorest region of the world, where plague, cholera, squalor and banditry reigned. Toncic continued: "agriculture and industry are foreign words here, people of this country make a living directly or indirectly through pilgrims".[15] Marschall concurred, saying "the Hejaz railway is completely worthless in economic terms".[16]

On top of everything else, except for the hajj season, there was no traffic density in the region—yet another reason why there could be no economic benefit associated with the railway, even in the long run. The administrators did not even bother to include relatively densely-populated areas such as Al Karak, Shavbak, and Mafraq in the railway.[17] Not surprisingly, there is not a single reference to the economic dimension of the Hejaz railway in Sultan Abdülhamid's memoirs. On the contrary, the military and religious significance of the line is accentuated:

> What is important for us is to be able to construct the railway between Damascus and Mecca promptly; thus making rapid soldier deployment possible in times of turmoil. A second important goal is to strengthen the bond amongst the Muslims to such an extent that would pound the British malice and deceit like a hard rock.[18]

As one can understand from the sultan's memoirs, the Hejaz railway was associated with religious and military/administrative factors. Before moving further, we should discuss the goals and limitations of Abdülhamid's Islamist policy, which he was trying to pursue in line with the Ottoman legacy.

2- *Abdülhamid's Islamist Policy*

A significant portion of the Balkan territories were lost in the aftermath of the war with Russia. In addition to the territories lost, 5.5 million Christian subjects remained outside the borders of the empire, which led to a significant change in the composition of the population. Henceforth, over 20 million of the 25 million people living within the borders of the empire were Muslims.[19] That is to say, in the 1880s, the Ottoman Empire was more of an Asian and Muslim state. Arab territories gained particular importance.[20]

Under those circumstances *Tanzimat* policy, which had relied on the European and Christian *millets* of the empire, was no longer justifiable. The Ottomanism of Tanzimat, which considered all subjects to be of equal status, left Muslims unsatisfied, as they no longer felt superior. The Ottomanism of the New Ottomans was a belated ideology which could not prevent the empire's disintegration. The goal of the New Ottomans was to establish a western-style constitutional regime. Abdülhamid, on the other hand, favored an absolutist/centralist regime which would hold the remaining parts of the empire together. Abdülhamid espoused a caliph-centered autocratic system instead of one based on a western-style social contract with his subjects. He was already aware that the majority of Muslim subjects were indifferent towards the parliamentary regime. For this reason he dissolved the first parliament at the first possible opportunity. The dismissal of constitutionalists such as Mithat Pasha and Namık Kemal marked the beginning of Abdülhamid's absolutist regime. In other words, the days that the Sublime Porte had had an important say were drawing to a close and the sultan was gaining absolute control again.

Abdülhamid's Islamist policy can be defined as a policy of New Ottomanism emblazoned with the ideological concepts of Islam. In one sense, secular Ottomanism was replaced by Islamic Ottomanism. This new type of Ottomanism was a highly pragmatic policy which legitimized an autocratic regime, capitalized on Islamic symbols, and highlighted the Islamic identity of the state.[21] Abdülhamid seemed to hold onto the Asian and Muslim identity of the state. The primary motivation of Abdülhamid's policy was to retain control over regions inhabited by Muslims. The sultan compared the empire to a tall plane tree. Loss of the Balkan territories meant getting rid of the blighted leaves. However, the trunk constituted by Islamic countries had to be preserved at all cost.[22] German military consultant Goltz Pasha, whose ideas were appreciated by the sultan, adopted a similar approach and stated that reinforcing the bonds between the Arabs and the caliph was more significant for the empire than losing a portion of Macedonia.[23] Under these circumstances, the trunk could only be preserved by pursuing an Islamist policy. In line with this policy, closer ties were established with such religious orders as the Rufai and Kadiri, religious figures were publicly recognized, new mosques were opened, a great number of religious books were published, religious festivals were celebrated more fervently, and the sale of alcohol was prohibited in Muslim neighborhoods.[24] Abdülhamid's practices that favored an Islamic lifestyle gave confidence and hope to Muslim people who had protested against the pro-western approach of the *Tanzimat* era.

It would be misleading to interpret the popular reaction to the westernism of the *Tanzimat* era in terms of a progressivism/reactionism dichotomy. At the root of this reaction was the extremely rapid socioeconomic metamorphosis that the Ottoman society was undergoing. It was evident even then that Westernism was not a project of change that the Ottoman society embraced through a process of internal evolution.[25] Due to its failure to defend itself even against internal uprisings, the government had turned to Europe in order to ensure its own territorial integrity—but this came at the expense of granting new capitulations. The articulation of the Ottoman market with the capitalist center through trade agreements had upset income distribution in an unprecedented manner. Artisans and peasants who could no longer survive against the competition coming from Europe had begun to live in misery, whereas pro-westernization *Tanzimat* bureaucrats had become even wealthier, which was the main reason behind popular resistance.[26] Those who were displeased with the new state of affairs were joined by *ulema* and landed proprietors and local notables who were excluded from governance due to renewed centralist policies. Muslim immigrants flocking into Ottoman territory from the Crimea, the Caucasus and the Balkans, feeling frustrated and degraded, were another important dynamic. Euro-centric policies resulted in the non-muslim comprador bourgeoisie getting richer and richer by acting as the commission brokers of Europeans. The anti-imperialist reactions against this group of bourgeoisie were another crucial factor that bolstered traditionalist/proislamist front. Apparently, when Abdül-hamid ascended to power, a disgruntled and traditionalist front had already been formed which was to constitute the backbone of his proislamist policy against westernist bureaucrat intellectuals.[27]

Islam had a very important place in Abdülhamid's foreign policy. Relations with Muslim countries were intense; religious leaders were dispatched to Algeria, Egypt, India and China to propagate the influence of the caliph. Moreover, newspapers published by Muslims were subsidized. Abdülhamid was using the religious orders to reinforce the support of the Islamic world. Religious orders, being one of the main components of social organization in Islamic countries, experienced a period of revival between 1880 and 1908. Their prominent leaders were gathered in İstanbul and were granted certain privileges. Thus these orders became effective instruments of communication and propaganda at Abdülhamid's immediate disposal.[28] Sheiks, seyyids and dervishes were sent to Asia and different parts of the world for propagandist purposes.[29] The elements of foreign policy based on Islamic union were as follows: Tartar, Mongolian, Georgian, and Circassian immigrants were warmly welcome, pilgrims from Kokand, Khiva, Bukhara were granted privileges to win their blessing; the great Arab thinker Sheik Abu'l Huda, and leaders of religious orders such as Sheik Rahmadullah, Sayyid Husayn al-Cisr and Muhammad Zafir were empowered to counterbalance the British and French influence. Abdülhamid was promoting his efforts by using the Rufai and Kadiri religious orders, and the caliph's name was pronounced at Friday sermons in Islamic countries. Briefly stated, the idea was to render Yıldız the Vatican of the Islamic world.[30]

The Europeans interpreted Abdülhamid's relations with the Muslim world as pan-Islamism and felt threatened. Particularly in the aftermath of the occupation of Tunisia and Egypt, the Ottoman administration was beleaguered by British and French accusations of pan-Islamism. European orientalists were producing a glut of books, brochures, and articles underlining the fear that Muslims would start an İstanbul-based insurgency and disrupt the colonial order. At the heart of this fear lay the fact that a sizeable Muslim population lived in these colonies. In fact only 20 million of 300 million Muslims lived within the borders of the Ottoman Empire. The remaining colonized Muslim peoples inevitably turned towards the Ottoman Empire as the only independent Islamic country. Muslims living outside the Ottoman territory had far-fetched expectations of Abdülhamid's administration—mainly because they had overrated the Ottoman Empire and the caliphate.[31] That was the main source of fear for the British and French as colonizers of Islamic countries. The same concern also troubled Russia, which had colonized the Muslim peoples of Central Asia.[32]

The Ottoman Empire was indeed the only independent Islamic country in the eyes of Muslim colonial peoples and their administrators.[33] Many Muslim countries, aspiring to regain independence, had enlisted help from the Ottoman Empire, and some had declared their allegiance to the caliph. During the rule of Abdülaziz (Abdülhamid's predecessor) Muslim rulers of Samara, Java, and the Comoros Islands had already written letters to İstanbul expressing their eagerness to create a common front under the leadership of the caliph. The relation of Algerian and Egyptian émigrés in İstanbul with the uprisings in Algeria and Tunisia had exacerbated French misgivings about pan-Islamism.[34]

The French Foreign Ministry was in the habit of requesting regular, detailed reports from its consulates in Muslim countries concerning the pan-Islamist activities of the caliph and the role of religious orders. One of the reports sent from Tarabya read, "the impact of panislamist propaganda has to do with Muslim nations' belief that the sultan is almighty". The consulate of Tripoli was asked to provide information about the activities of the *Shazaliya-Madaniya* order, their relationship with the caliph, their influence on the community, and about how to "weaken the pan-Islamist movement". The report stated that "the orders were used to intercept France's infiltration into Africa". The report continued as follows:

> Arabs venerate and submit to him [Abdülhamid] because of his being the caliph . . . The Sublime Porte, since taking control of Libya . . . managed to keep this spirit alive within the Arabs and used influential local figures for this purpose . . . granted those people titles. . . and it is fair to say that these figures have the power to enlist help literally from all Arabs and Africans. In a nutshell, the Ottoman government is using Islamic fanaticism as raw material to produce a weapon against Europe and trying to render this weapon unbeatable.[35]

The radical writings of truly pan-Islamist writers such as Jamal ad-Din al-Afghani demonstrated that the imperialists' misgivings about pan-Islamism were not misplaced. Not only France but also Britain, terrified by the Muslim uprisings of 1857, was concerned that Abdülhamid was eager to unite all Muslims under the aegis of the caliph. Russia, with all the Muslim khanates of Central Asia under its control, felt the same way. However Abdülhamid knew the limits of his power, and thus pursued a strictly defined policy of preserving the status quo. Therefore he could not plausibly be expected to try and forge a political union with all Muslims of the world under his leadership. Abdülhamid was capitalizing on the pan-Islamist phobia of the Europeans as a potential threat, at best. We know that during his 33-year rule, Abdülhamid always avoided adventureous politics. He was only interested in gaining the prestige of the spiritual leader of the Islamic world; in fact, he was interning the genuine pan-Islamist who he feared could manipulate him. He had lured Mizancı Murat and Jamal ad-Din al-Afghani to İstanbul by giving false promises and then neutralized them. When the two thinkers, seduced by the favors granted by the sultan, understood that they had been trapped in a golden cage it was already too late.[36]

Interestingly, despite Abdülhamid's cautious policy, the Germans were strongly in favor of pan-Islamism. German attitudes, expectations and statements were provocative and alarming, especially with respect to the British. Emperor Wilhelm, in his second visit to Abdülhamid, had proclaimed himself to be the guardian of 300 million Muslims. Renowned orientalist Max von Oppenheim wrote an official report which illustrated that the Germans had taken their pan-Islamist ideas even farther than had Abdülhamid. According to Oppenheim, India, Egypt and other British colonies were going through a process of self-awakening. The Islamic world had become a nightmare for the British. Muslims had turned to the caliph, the most respected Islamic leader, and the sultan/caliph was therefore obligated to realize this potential. "A less cowardly sultan" harnessing the immense power of the Islamic world could bring Britain to her knees. Britain, terrorized by the power of Islam, could never venture to go to war in Europe, particularly against Germany. Therefore Germany was trying both to secure close ties with the sultan and to convince him to wield the power of Islam. Britain, aware of what was going on, was striving to take over the holy cities and the office of the caliphate from the Ottomans. Had Britain succeeded in this, nothing could have deterred her from carrying out a policy of aggression in the Middle East.[37] Pan-Islamism was an everlasting dream for Germany. Kaiser Wilhelm kept pursuing this dream throughout the years of World War I as well.[38] Enver Pasha saw eye to eye with the German general staff. Agents of the clandestine organization *Teşkilat-ı Mahsusa* tried hard to spark rebellions in the Allied powers' Muslim colonies.[39] All their plans, however, fell through.

What was passing through Abdülhamid's mind? What was the goal of establishing ties with Muslim peoples based on the common denominator of Islam and the supremacy of the caliphate? Goltz Pasha, the long-time mission commander for the German military delegation in the Ottoman army, analyzed the situation as follows: "Ottomans were trying to compensate for the loss of reputation which resulted from increasing Christian/European influence at their doorstep by boosting their prestige in the Islamic world; the purpose was 'to conquer from within', so to speak."[40] According to Bodman, a German diplomat, this policy was the best cure for the frustration caused by loss of territory and morale.[41] As a matter of fact, as Goltz Pasha stressed, Abdülhamid was not pursuing an expansionist policy, his was a policy of defense, aiming to mobilize Muslim peoples within his borders under the leadership of the caliph. If there was pan-Islamism of any sort, it was an "internal pan-Islamism".

Put differently, Ottoman pan-Islamism was the policy of a state which felt threatened and wanted to defend itself. The goal was to reestablish the legitimacy of the Ottoman sultan as the caliph of the Muslim world and use it as a trump card against imperialist countries.[42] Garnering support from Muslims abroad would strengthen his authority over Muslims at home. In a nutshell, Abdülhamid's role as the caliph and leader of all Muslims resulted from the dialectic interaction of three different developments: the attitude of colonized Islamic countries towards the West and the Ottoman Empire; the concerns of colonizers due to the fact that the majority of their colonial subjects were Muslims; and Abdülhamid's willingness to capitalize on these two sentiments by wielding the title of the caliph to assert his legitimacy. One of the two types of sentiment was that, the colonizers were haunted by the ghost of pan-Islamism that they themselves had created.[43] The other sentiment was that, Muslim peoples living in those colonies had put too much trust in the Ottoman Empire and the office of the caliphate. Abdülhamid was trying to pursue a balanced policy: he was striving to benefit as much as possible from the use of an imagined pan-Islamism avoiding at the same time any risks involved.

As the virus of nationalism kept infecting Christian elements of the empire, the composition of the population changed in favor of the Muslims, rendering four-fifths of the Ottoman population Muslim. Ottomanism could no longer act as the ideological mortar that held different elements of the empire together. Given that the majority of the people were Muslims, this mortar could only be Islamism. Abdülhamid pronounced that the social structure and politics of the

Ottoman Empire was based on religion and that "racial origin" should never be a point of concern, everyone had to embrace the state with their "Muslim" and "Ottoman" identity.[44] The strategy henceforth had to focus on preventing the virus of nationalism from infecting non-Turkish Muslim peoples, such as the Arabs, Kurds and Albanians. Abdülhamid, aware of the danger, had dissolved the Albanian League of Prizren and garnered support from Albanian and Kurdish tribal leaders.

The greatest threat was a possible separatist movement in the Arab provinces. In the aftermath of the 1877–78 defeat, the British had concluded that the Ottoman Empire could not be relied on to bar the road to India to the Russians. New British foreign policy would focus on controlling strategic locations to ensure the safety of the road to India and building an "Eastern Security Road" between Britain and India. This road would run from west to east and link strategic spots such as Gibraltar, Malta, Cyprus, Egypt, Aden, and the Persian Gulf.[45] Britain, enjoying control over Gibraltar, Malta and Aden, would rapidly secure its domination in Cyprus and Egypt, foster ties with the "sheiks" alongside the Persian Gulf and create an area of influence.[46] This explains why the Suez–Red Sea shores and the Arabian peninsula gained particular importance for Britain. Abdülhamid understood that the British would play the Arab card to seize control of strategic locations within the empire. There were already signals that Arabs would have a greater role to play in the new British foreign policy.

As far back as 1839, after the conquest of Aden, the British started to spread their influence all the way down to Yemen. They were supporting the rebellious Yemeni tribes with weaponry and money. For instance, Abdali Sheikh Ahmad Fodol joined the British camp in return for 30,000 rupees.[47] It was obvious that the British would pursue the same policy in the Hejaz as they did in Aden, Nejd and Yemen[48]. The events unfolded as expected. Arab newspapers published in London had launched a pro-Arab caliph campaign in the aftermath of the Russo–Turkish War. The main idea was to restore the office of the caliphate, which the Ottomans had usurped by force, to its rightful owner, the Arabs. For example, on October 19, 1876, the *Mirat al-Ahwal* newspaper, published in London with British funding, declared the Ottoman caliphate invalid and advocated an Arab caliph.[49] Likewise, Wilfrid Scawen Blunt, a British writer and poet, spoke for the same arguments in his writings.[50]

The British were trying to restore the office of the caliphate to Egypt, surmising that, should they succeed in transferring the office of the caliphate to Egypt, which was under their control, they would have a better chance of reaching out to the Islamic world. Abdülhamid was fully aware of Britain's intentions and said "in order to undermine the influence of Islam and strengthen their dominance, they want to crown the Khedive as the caliph". The sultan was also sarcastic about the British advocacy of an Egyptian caliph: "It would not come as a surprise to see the British Lord Cromer crowned as caliph if need be."[51]

British propaganda began to spread in the Arab provinces of the Ottoman Empire. Starting from the 1880s, they distributed manifestos in Beirut, Aleppo, Damascus and Baghdad inviting people to get rid of the Ottoman yoke. A pro-British wind had started blowing. In 1902, the Syrian al-Kawakibi's book *the Mother of Cities* was published. The author was describing an Islamic revivalism project which would spring from Mecca, and which advocated an Arab caliph. A few years later another author, Nagib Azoury, wrote a book in French, *The Revival of the Arab Nation,* and openly propagated Arab nationalism. The author also wrote an article in Paris calling upon Muslim and Christian Arabs to unite under an independent Arab state.[52] Christian Arabs like Ibrahim al-Yaziji had started to advocate a secular form of nationalism.[53] For Yaziji and others, being an Arab was superior to being a Christian and they were determined to join forces with their Muslim brothers.[54]

In 1900 British newspapers were openly campaigning for British domination in Arabia. In his memoirs Abdülhamid wrote: "British newspapers have often been vocal about Britain's Arabia policy. However none of them had ever argued that Arabia should concede British dominance or that Britain with 56 million Muslims should control the holy cities."[55] Arabia had a special significance for the sultan due the holy cities of Mecca and Medina—not surprising, given that Abdülhamid was pursuing an Islamic policy with the caliph at its center. Just like many other German ideologues, the orientalist Roloff accentuated this fact. In Roloff's words, if the control of the holy cities had been taken over by Arabs who were under British protection, this would have meant "the political demise and the end of Turkey".[56]

Ottoman sultans bore the title of *hadim el-haremeyn el-şerifeyn*, which meant "servants of the two holy cities", Mecca and Medina. This title was a major source of legitimacy for the sultan as the guardian of hajj. A caliph who lost control of the holy cities would lose the credibility of his Islamic policy as well. If one pillar of Abdülhamid's Islamic policy was the office of the caliphate, the other one was the organization of hajj. The hajj functioned as the general assembly of Muslims from all over the world. In fact, Abdülhamid was eager to do whatever it took to win the hearts and minds of the Muslims during hajj season. The hajjis who got sick were treated in the hospitals constructed by the caliph. Those who ran out of money were subsidized. People of influence from Asia and India were welcomed with special ceremonies and all their needs were duly satisfied.[57]

The region that harbored the holy cities was integrated with the empire through political rather than military mechanisms. Because of its distance from the capital and its unique social structure, the Ottomans had a limited number of troops deployed in the Hejaz. Therefore the nexus of interests which tied different segments of the Hejaz community to the state had to be maintained. The amirs were articulated with the Ottoman system through *atiye-i hümayun*, Bedouins through *surra* allowances, inhabitants of Mecca and Medina through religious endowments and official funds. That is why the departure of the *surra* regiment was a meticulously organized ceremony. Now that the amirs, the Bedouin sheiks and prominent inhabitants of Mecca had accepted "the benefits granted by the sultan", they had to recognize his legitimacy as well.[58] Therefore the sultan had two choices before him: he could either maintain the system of integration already in place or replace it with other mechanisms. We will soon discuss the role the Hejaz railway played in this respect. For the time being, suffice it to say that the probability of losing Arabia was Abdülhamid's greatest phobia.[59]

Abdülhamid was planning to take countermeasures against the British efforts to sever Arabia from the empire and trying to formulate effective policies. One of those was supporting propagandist publications in Arabic which advocated the legality of the Ottoman caliphate and hence discredited an Arab caliph. For instance, as a response to the *al-Khalifa* newspaper published in London by a priest of Syrian origin, Louis Sabunji, Abdülhamid had supported the publication of *al-Ghayrat* newspaper. The supporters of *al- Khalifa* included not only the British but also Khedive Ismail. In response to Sabunji's claims that the Ottoman caliphate was fabricated, and that the Ottomans were oppressing the Arabs, *al-Ghayrat* campaigned for the Ottoman caliphate. This newspaper, funded by Abdülhamid, was published by an Indian journalist, Mawlanah Abdul Resul.[60] Another newspaper advocating an Arab caliph was published by Ibrahim Muwaylihi in Naples. Interestingly, Muwaylihi's newspaper was also named *al-Khalifa*; both Sabunji and Muwaylihi were later bought off by Abdülhamid and submitted to his authority.[61]

Naturally Abdülhamid suppressed the voices of those who claimed that caliphate should derive from the Quraysh tribe. Classical Islamic books on politics attributing the office of the caliphate to the Quraysh tribe were banned. Likewise books written by some modernist Islamist authors, such as the Indian Sayyid Ahmad Khan,[62] who asserted the same, were intercepted at the Ottoman borders. The jurist and political theorist Cevdet Pasha pointed out that these opinions were provoked by the British and tried to legitimatize the Ottoman rule and caliphate by saying, "Islam obliges one to submit to power and might."[63] Similarly Lutfi Pasha stated that the leader of the greatest Islamic state who was powerful enough to protect the religion and sharia was also entitled to become the caliph.[64]

Now that the Arab provinces had gained priviledged status in the eyes of Abdülhamid, the resources allocated to the region also soared; Arab cities such as Beirut and Damascus were given priority in terms of education and infrastructure investments. Morever, Abdülhamid dispatched his most trusted and talented administrators to the region. Arabs had started to play bigger roles in the state apparatus as well. Arabs who occupied prominent offices included Deputy Secretary of the Palace Izzat Pasha, Lebanese Maronite Selim Melhame Pasha, Abu'l Huda al Sayyadi from Aleppo, Muhammad Zafir from Tunisia, and others. Abdülhamid had always respected and admired Arab thinkers. Thus he convened a group of administrators who were aloof from the power struggles in İstanbul and had declared allegiance to the sultan. Arab involvement in Ottoman bureaucracy and administration gave impetus to his centralizing/proislamist policy. The number of Arab officers serving in the Ottoman army increased to 3,200 in 1886. Until then Arabs had had a negligible presence in the higher echolons of Ottoman administration. None of the 215 Ottoman grand viziers were of Arab origin. There was not a single Arab navy chief. There was one Arab among the head treasurers (*başdefterdar*) and four among the heads of the secretary and Foreign Office (*reisülküttap*). One of the complaints raised in a communique issued by the Association for the Protection of the Rights of the Arab Nation concerned the fact that Arabs were not promoted as high-ranking military officers. This shows that Abdülhamid's approach was indeed realistic.[65]

The sultan had decided to address the Arab tribes by using Islamist language and harnessing the influence of the office of the caliphate. He was, however, aware that he could not implement this policy by referring to laws or resorting to force. The dominant lifestyle in this region, where state authority could never be established, was nomadic. Arab sheiks and amirs played crucial roles in the social, economic, religious and political lives of Arab tribal leaders. Therefore Abdülhamid implemented his policy by forging personal ties with Arab tribal leaders. As a matter of fact, the caliph relied on prominent members of strong families from Damascus or Aleppo or tribal leaders for his legitimacy in Arab provinces. He would win the allegiance of those leaders by honoring them with gifts, concessions and compliments.[66]

In order to intensify the relationships with the tribal leaders, Abdülhamid inaugurated *Aşiret Mektebi* (the School for Tribes) in İstanbul in 1892. The idea was to educate and mould the sons of prominent sheiks and tribal leaders in line with the sultan's objectives. In other words, to inculcate the tribal leaders of the future with Ottoman culture, teach them Turkish and ensure their allegiance to the state, the sultan and the caliph.[67] The decision to establish this school was influenced by the enrollment of 48 students from Arab tribes in Harbiye (Military Academy) and the satisfaction it created among the Arabs.[68]

The statute of the School for Tribes stated that students from the Hejaz, Syria, Aleppo, Baghdad, Basra, Mosul, Tripoli, Yemen and Diyarbekir would be admitted. Children of "venerable and respectable" families who were between the ages of 12 and 16 and "mentally and bodily" healthy, were eligible candidates for this boarding school. The students would receive a monthy salary of 30 kurush.[69] The students were invited for *iftar* twice during the holy month of Ramadan by Abdülhamid.[70] The alumni were supposed to teach at schools which were to be opened in their tribal hometowns or to assume an administrative office. Article nine of the statute unequivocally expressed the raison d'être of the school as ensuring and strengthening the loyalty of Arab tribes to the state.[71] Oppenheim pinpointed the potential role the school could play in instigating the Bedouins to settle.[72] The school was clearly an offshoot of Abdülhamid's policy of holding Muslim elements of the empire together and thus ensuring an Islamic union. Abdülhamid's administration, discomfited by Arab intellectuals' demands to use Arabic in education and administration, was hoping to educate their children and assimilate them within the bureaucratic apparatus. Between 1902 and 1907 a total of 167 Arab students were enrolled in *Mekteb-i Mülkiyye* (School of Administration).[73]

In line with the same policy, Abdülhamid was trying to increase the number of Turkish schools in the Arab provinces to counterbalance the number of foreign schools. In the 1880s, there was one *idadi* (primary school) and 20 *rüşdiye* (middle schools) in Aleppo and one *idadi* and three *rüşdiye* in Damascus and their number was on the rise. Although the language of the press in the region was predominantly Arabic and French, six newspapers were published in Turkish. Religious leaders, such as Rufai sheikh Abu'l Huda of Aleppo, were brought into the fold.[74] All of these were attempts to tame the inevitable reality of Arab nationalism with official Ottoman centralism.

3- *Religious/Political Factors*

Within this general framework, the Hejaz railway was a very important pillar of Abdülhamid's Arabia policy, a gigantic step towards increasing the chance of success of this highly shaky policy. The proclaimed goals that the Abdülhamid administration was aiming to achieve with the railway were to connect the holy cities with Damascus, facilitating the otherwise very dangerous and challenging hajj journey, and to increase the number of hajjis and promote Islamic solidarity. These goals added to the effectiveness and prestige of Abdülhamid's Islamist policy. Nevertheless, although these were sincere statements, this initiative obviously had other unannounced objectives/ulterior motives such as blocking British separatist propaganda in the Arab states and preventing the secession of the holy cities. Put differently, Abdülhamid wanted to avoid the transformation of the holy cities into an Arab state ruled by an Arab sheik who would also hold the title of caliph. This is also reaffirmed in Abdülhamid's memoirs, where he explained the purpose of the Hejaz railway as rapid deployment of soldiers to the region and "strengthening the bonds amongst the Muslims to such an extext that British treason and fraud would crash against that hard rock and crumble".[75]

The Hejaz railway project was indeed a very useful instrument to "strengthen the bonds amongst the Muslims". By constructing the railway, the sultan/caliph would be able to consolidate his position as the religious leader of all Muslims, not only those living in Ottoman territory. Given that 15,000 hajjis from India and 80,000 from Iran alone visited the holy cities[76] and that one-fifth of those lost their lives during the ardous journey,[77] Abdülhamid's reckoning was justified. For, since the railway would have rendered the perilous hajj journey easier and cheaper, it would have earned Abdülhamid an enormous amount of respect in the Islamic world.

Without the railway, the hajj was indeed very challenging and risky. The high death toll is self-explanatory. Due to the lack of settled areas en route, food shortage posed a serious problem. Nevertheless the biggest threat was the lack of water and extreme heat while marching through the desert. The water shortage became unbearable when the temperature rose to 50–60 degrees celsius and all springs and wells dried up. Apart from being in short supply, the available water was not clean enough, and that led to typhus and cholera outbreaks among the pilgrims.[78] The hajj season would shift during the year according to the Islamic calendar; therefore extreme low temperatures could also pose serious threats in the midst of the desert. In 1846, for instance, 1,200 horses, 900 camels and 500 hajjis in a Syrian caravan froze to death. Inadequate environmental and health conditions, overcrowding and malnutrition were additional risk factors for the pilgrims.[79] Hajjis could not survive the diseases they caught in the holy cities since their bodies became feeble during the challenging journey.[80]

On top of all these challenges, one must add Bedouins, who would waylay and rob the caravans. Seen in this light, it required a lot of courage to embark on this journey in the first place.[81] Whenever the distribution of the *urban surra* was delayed, Bedouin tribes protested and plundered the hajj caravans. In addition to all the dangers and problems associated with it, hajj was also a very expensive undertaking. An average hajji had to spend at least 50 lira for that part of the journey between Damascus and Mecca.[82] Still, some poor hajjis were able to reach Mecca by serving well-off hajjis on the caravan or by travelling on foot, which rendered an already toilsome journey even tougher. Traveling from other countries to Damascus was very costly as well.

Therefore a lot of Muslims in different parts of the world could not afford to go on the hajj, much as they might wish to. This explains the importance of Abdülhamid's Hejaz railway project for the Muslims. The railway would spare thousands of hajjis this costly and risky journey. Journey by railway was not only safe and comfortable but could be completed in eight days. However, travelling from Damascus to Mecca took around 50 days for the hajj caravan. Constructing a connection line to Jeddah would facilitate access to the holy cities following a sea voyage.

The Ottoman sultans had been responsible for the security of the hajj caravans traveling to and from Mecca since the 16th century. Any problem—a Bedouin attack, or trouble in supplying food or water— would imply a crisis of legitimacy for the sultan. The orderly conduct of the hajj could only be maintained if the Ottoman administration controlled central and local elements on the hajj route in a balanced manner. Hajj was a complicated journey which took more than three months in total. Escorting the Syrian caravan of 20–60,000 hajjis was such a crucial responsibility that since 1708 the governor of Damascus had been entrusted with this duty. Moreover, hajj organization was very costly for the Ottoman Empire. Through the end of the 19th century, donations and allowances allocated to the Bedouins for the travel expenses and supplies of a thousand soldiers would cost approximately 12,500,000 para.[83]

The organization of the hajj was obviously a complicated process: wayward amirs of Mecca had to be contained by Ottoman governors in Jedda (they received imperial allowances on a regular basis); Bedouins had to be prevented from attacking the caravans by paying them *surra* (in exchange for which they were beneficial in supplying food, water and camels); the cooperation of the inhabitants of Mecca and Medina had to be ensured through a free supply of grain. Such measures could be effective only if they were monitored closely. There were so many different sociopolitical groups with overlapping domains of activities that a minor failure could bring the entire machine to a halt. For instance, failure to dispatch the *surra* to be paid to a tribe on the desert road in time could put the lives of hajjis in danger.[84] Massive casualties among the Bedouins or their camels due to drought or infectious diseases could lead to a similar outcome. This whole system could only run provided that the Ottoman Empire kept pouring all kinds of external resources into the Hejaz and Red Sea regions. Numerous foundations established across the country served this very important function.[85] Ensuring the safety of the hajj journey was so critical for Ottoman sultans that even in times of financial distress, the payments were made. Briefly stated, Ottoman sultans, bearing the title of *servants of the two Holy Cities* had always been concerned about maintaining law and order during the hajj. Abdülhamid assumed that the Hejaz railway would earn him prestige in the Islamic world and solve all the above problems at the same time.

Due to its significance for the Muslims, the Hejaz railway project was welcomed with a lot of enthusiasm by the Islamic world. The Ottoman press, when covering news about the Hejaz line, would make far-reaching claims about its potential benefits to the Ottoman Empire and the Islamic world. Ottoman newspapers such as *Sabah* and *İkdam* invited Muslims to support the Ottoman government in realizing this "auspicious" project. The Hejaz railway was a headline story in the Indian, Iranian and Arab press as well. For example, *al-Muayyad*, published in Egypt, heralded the project with gratitude and stated that the caliph would be blessed everywhere because of his determination to proceed with the project. Similarly, the Islamic newspapers published in India (*al-Wakil*, *al-Watan* and the like) praised the Hejaz railway and called on Indian Muslims for help with the construction of this holy line. The *Muhammedan* newspaper wrote that the Hejaz railway was the single most important undertaking in the history of the Ottoman Empire and therefore all Muslims should be grateful to this great Islamic state.[86] The Hejaz railway had created excitement in the entire Islamic world from Morocco to Egypt, from Russia to China, and Muslims were called upon to support the caliph's project.[87] The railway was dubbed the "holy line", the "magnum opus of the caliph", the "immortal work". Abdülhamid was depicted as the "religious leader who was working tirelessly to satisfy the requirements of the Muslims". In the eyes of Muslims, Abdülhamid and the Hejaz railway were one and the same. In fact the official name of the railway was *Hamidiye Hejaz Railway*.

In short, thanks to the Hejaz railway initiative, Abdülhamid garnered unprecedented support from the Muslim community and boosted his influence and reputation as caliph. His success was a rebuke to British-backed propaganda for an Arab caliphate. A journalist summed up the status of Abdülhamid as follows: "He became at one shot the most popular man of the Islamic world, and hero of the day."[88]

4- *Military/Political Factors*

As seen above, Abdülhamid's reputation within the Islamic world had risen as soon as he announced his decision to construct the Hejaz railway. However, the sustainability of this success depended on maintaining his military and political domination in the Hejaz region. Holding sway over the Hejaz region required rapid soldier deployment. The most reliable solution would be to construct a railway from Damascus to Mecca. That is to say, by embarking on the Hejaz railway Abdülhamid would be killing two birds with one stone: it would give him the opportunity to implement his islamist policy and also the hope of strengthening his authority in the region.

The British seizure of the Suez Canal had increased the military and strategic significance of the Hejaz railway for the Ottomans. The canal ensured British domination over the Red Sea coast, Hejaz and Yemen; the British-controlled canal was the only route by which the Ottomans could despatch soldiers and ammunition to the Hejaz and Yemen. A possible war with Britain would have led to blocking of the canal for soldier transfer. Thus Abdülhamid reemphasized the significance of the Hejaz railway in his memoirs: "Once the road to Mecca is completed, we will be able to deploy our soldiers safe and sound, hence the Suez Canal will no longer be required."[89] In addition, construction of a connection line to the Aqaba Gulf would have enabled access to the Red Sea, hence eliminating the need to use the canal altogether.[90] Many Ottoman administrators, such as Extraordinary Commisar to Egypt Ahmet Muhtar Pasha, had raised the importance of the Aqaba railway repeatedly.[91]

Arabia had been a part of the Ottoman territory since the 16th century. Nevertheless the Ottomans could never enjoy complete domination over the Arabian peninsula. It was impossible to rein in the freewheeling Arab tribes who were well accustomed to extreme desert conditions. They could neither collect taxes nor recruit soldiers from the Arab tribes. Moreover Arab Bedouins, as part of their nomadic lifestyle, would loot and plunder quite often. Archives of the *qadi* frequently mentioned Bedouin attacks on travelers, the pressure they put on the peasants and the peasant uprisings they provoked.[92] For instance, in 1757, when the news that the Syrian hajj caravan had been plundered reached Damascus, a major crisis broke out. In 1900 not much had changed.[93] There was no stable military force to deter the Arab nomads from plundering; likewise, the socio-economic conditions, which could have instigated the nomads to settle, were also not there. The state had to shake hands with Bedouin sheiks in order to contain them. It has already been mentioned that the Bedouin received *urban surra* in return for non-agression. Another method used to stop Bedouin attacks was to recruit them as guardians of the hajj caravans. [94]

Thus the Ottoman Empire found it harder to establish its authority in an increasingly important region. Ottoman domination, which had already been far from strong, was jeopardized by British provocations. Ottoman troops found it hard to quell recurrent uprisings. Riyad Sheik Ibn Saud was aspiring to found a Wahhabi Empire by rallying forces with prominent sheiks of the Arab peninsula.[95] Commander of the Fourth Corps Marshal Fevzi Pasha, who was sent to confront Ibn Saud, was blockaded in Sanaa.[96] Ottoman archives prove that Ibn Saud provoked the soldiers who had been deployed to quash the uprising in Yemen.[97]

To make matters worse, the amirs of Mecca were a law unto themselves. The amir sheld the religious and worldly leadership of the Hejaz. The most influential authority in the region before Ottoman domination, they remained so after the Ottomans had gained control. Such was their power that, in addition to *atiye-i hümayun*, the Sublime Porte had to send gifts to please the amirs. Moreover, all earnings from the hajj were transferred to the amirs. For instance, the Bedouins had to give a considerable amount of their earnings from camel transportation to the amirs. Considering their influence on the Arab tribes, the amirs' waywardness obviously created an additional problem for the Ottoman administration.[98]

Therefore, Abdülhamid had been looking for ways to tighten military control over the Arab provinces. Expert reports validated his strategy. A railway to connect Arabia with Syria would ensure rapid and safe soldier deployment without having to use the Suez Canal, which was under British control. Inauguration of the railway would bring Bedouin uprisings to a halt and pave the way for military and administrative measures in the province of Arabia. As a matter of fact, one year after the decision to construct the Hejaz railway had been made, the Guardian of Hajj (Amirul Hajj) Muhammad bin Abdul Hamid reported that certain Arab tribes on the hajj road parted ways with Sheik Abdul Aziz al Rashid and declared allegiance to Ottoman authority.[99] In addition to military measures, steps needed to be taken to boost economic welfare in the region. This could best be done by tying the region to the imperial capital by rail.[100]

Abdülhamid had brought up the Hejaz railway project even before Baghdad railway negotiations were concluded, for Abdülhamid regarded these two projects as parts of a whole, not as independent undertakings. Once these two railways intersected in Aleppo, this could lead to a shift of authority in Arabia in favor of the Ottomans, connecting major parts of the Arabian peninsula such as Syria, the Hejaz region, and Baghdad with the imperial capital. Soon after announcing his Hejaz railway project, Abdülhamid told Kurt Zander, General Manager of the Anatolian Railway Company, that it was his heartfelt desire to see a railway built between the capital city of İstanbul and the holy city of Mecca. Consequently, he requested Zander to construct a connection line between Baghdad and the Hejaz railways as soon as possible.[101]

German Commander Auler Pasha stated that constructing a Bulgurlu–Aleppo section of the Baghdad railway would be enough to tie the two railways and boost their military importance. According to his calculations, a railway would reduce the travel time from İstanbul to Mecca to 120 hours. It would take six days for military troops to cover this distance.[102] During World War I, the Baghdad and Hejaz railways indeed played a very crucial role in dispatching soldiers to Palestine, as projected by Abdülhamid and Auler Pasha. The troops were sent to Aleppo by the Baghdad railway and then transferred to Medina via Damascus by the Hejaz railway.

C) THE EVOLUTION OF THE HEJAZ RAILWAY PROJECT

The idea of building a railway in the Hejaz had been raised many times since 1860. The 1860s were the heyday of railways. It was back then that Zimpel, an American citizen of German origin, proposed a railway project to connect Damascus and the Red Sea. That was the very first railway project put forward for the Hejaz Region. The French consul to Jeddah had stated at the time that the amir of Mecca did not favor any railway project in the region. The amir feared that the railway operation would disrupt his monopoly on camel transportation and hence take away his income.[103] Zimpel had also emphasized that Bedouins would oppose the project on the same grounds.

Colonel Ahmed Reşid Pasha was the first insider in the Ottoman administration to formulate a proposal concerning the Hejaz railway. Ahmed Reşid had participated in the Yemen expedition in 1871–73 and reported his observation in 1874. He had seen that it was extremely difficult to deploy Ottoman soldiers to Yemen and suggested that the only way to overcome this challenge was to build a railway from Damascus to Mecca and Jeddah. The Pasha also underlined the importance of the office of the caliphate and Arabia for the Ottoman Empire.[104] Osman Nuri Pasha was another Ottoman officer who had served in the Hejaz and concluded that the security of the region could only be maintained by constructing a railway in Syria, the Hejaz and Yemen. The Jeddah–Arafat railway project, formulated by Dr Kaimakam Şakir in 1890, aimed to serve another purpose, that of solving the healthcare problems of the hajjis, a complaint that had been used as an excuse to challenge the Ottoman authority.[105]

Two years after Kaimakam Şakir, the commander of the Hejaz military division, Haji Osman Nuri Pasha, suggested the construction of a railway between Jeddah and Mecca in order to "overcome the material and physical challenges and predicaments that hajjis suffered during their travel by camels" and to prevent Bedouin attacks. A local company with Muslim shareholders, engineers and manager, could have been established and a light railway could have been constructed without putting a strain on the Ottoman treasury.[106] According to the Pasha's remarks, the amirs of Mecca were the main opponents to a railway in the Hejaz. When he had suggested to Amir Awn al-Rafik and Hakkı Pasha, the governor of the Hejaz, that they draw up a joint report about a Jeddah–Mecca railway to be submitted to the sultan, both men had used different excuses to drag their feet. Nevertheless, a commission in İstanbul assessed Osman Nuri's proposal and deemed it eligible. The commission stated that the difficulties of hajjis would be overcome by the construction of the proposed railway.[107]

The Hejaz Itinerary submitted to the sultan by a young Ottoman officer, Süleyman Şefik, in 1891, was in the form of a comprehensive report concerning the desolateness of the Hejaz road, and the Arab tribes. He suggested constructing a 200-km railway that would lead from a harbor in Syria to the Aqaba Gulf.[108] One year after Şefik's proposal, in December, 1892, Muhammad Hasan Seltani al-Mahzuni applied for concession for a railway from Damascus to Mecca and Jeddah. He was representing a company named *Şirket-i Osmaniye ve İslamiye*. The Ottoman administration, taking into account a possible foreign capital involvement,[109] interrogated the applicant thoroughly. The investigation proved that the applicant was not affiliated with any foreign capital. He argued that wealthy Muslims would have done great service to the Islamic world by establishing an Islamic enterprise and building a railway to Mecca. The investigation report written by experts highlighted the main factors which successfully gave rise to the Hejaz railway idea and project [what follows is an abbreviated and abridged version of the text—MÖ]:

> ...unless an alternative way, other than the Suez Canal controlled by the British, is sought to connect the holy lands to the city of the caliph, the Red Sea Coast and the Hejaz region might fall pray to the evil intentions of those who strive to destroy the very foundations of the caliphate. Islamic peoples who are obliged to fulfill their hajj duty try to reach the holy cities either by foreign vessels where they are subjected to

humiliation or by camel, a journey through which they have to endure a lot of challenges including drought for months. Some bedouin are rewarded with gifts and allowances in return for ensuring the safety of the caravans, however we have recently seen that this practice has its shortcomings. It has become mandatory to construct a railway in this region, both to solve these problems and to show the power of the caliph. ... The railway has to be built solely by Muslim involvement. Therefore previously untested methods will have to be employed, such as obtaining a huge amount of finance from the Islamic world and recruiting Muslim engineers in its construction. The Military School has been teaching courses in railway science and many officers have graduated with expertise in this field. This might solve the problem of recruitment. However, 5 million liras will be required for the construction of approximately 1,700 kilometers of rail. It will obviously be a challenge to collect such a huge sum from Ottoman Muslims at once. Nevertheless, inviting all Muslims to join forces for this religious deed under the leadership of the caliph and extending the money collection over time will undoubtedly eliminate this problem ... In a nutshell, our sultan must personally lead this highly significant undertaking. Muslims hold our sultan in very high regard; therefore people of political and economic influence will not hesitate to allocate some of their assets to this cause when they see our sultan personally leading the initiative. State officials will also be eager to make all kinds of sacrifices for his lofty cause so it might be possible to collect a huge sum of money at the beginning.[110]

Izzat Pasha's was the most detailed of all Hejaz railway reports. Izzat Pasha, born in Damascus, was the son of a wealthy landlord of Arab origin. His father, Holo Pasha, was honored with titles and medals by Fuad Pasha. Cevdet Pasha, Minister of Justice, invited Izzat Bey to İstanbul and promoted him to the head of the Commercial Court.[111] Izzat spent his entire childhood listening to challenges and predicaments experienced during the hajj journey.

When Izzat submitted his report to the sultan in 1892, he was the Director General of Foundations in Jeddah. Then he was appointed to the Council of State, became the chamberlain,[112] and later on assumed the most prominent positions during the construction of the Hejaz railway.[113]

Izzat Pasha drew attention to the dangers associated with the closing of the Suez Canal by the British in the face of an imminent war. In order to maintain control of the holy lands, something had to be done. Izzat Pasha asserted that the Ottoman Empire had to continue being the guardian of Medina, the burial place of the prophet and Mecca, the *qibla* of all Muslims. Another important point that Izzat Pasha touched upon was the Ottoman rulers' failure to establish authority in the Hejaz region except during hajj season. If the Hejaz railway were built, deploying soldiers would cease to be a problem. The hajj would become less challenging, which in turn would boost the number of hajjis and hence the reputation of the sultan within the Muslim community. Apparently Izzat Pasha considered the Hejaz railway an important element of Abdülhamid's Islamist policy and an instrument of domination in the region. Besides, a rapid increase in the number of pilgrims would enhance economic opportunities. The soil was very infertile and the needs of the region were satisfied from other cities such as Damascus and Aleppo. It made economic sense to connect the Hejaz with these cities by rail.[114]

We have already elaborated on the Egypt–Syria, Egypt–Persian Gulf, and Jeddah–Persian Gulf projects formulated by the British. In 1897, partly as a reponse to these debates and partly as a reaction to increasing British domination in the region,[115] a Muslim Indian writer, Muhammad Inshaullah, proposed a railway project which would commence from Damascus and lead to Yemen via Medina and Mecca. Inshaullah had mentioned this project for the first time in his book on the Ottoman dynasty. He kept expressing his opinions through articles published in various newspapers, including *al-Wakil and al-Watan,* where he was working as editor-in-chief.[116] Inshallah had submitted articles to the İstanbul-based *al-Malûmat* and Eygpt-based *al-Muayyad* and *al-Manar* newspapers as well. Overall, these newspapers declared support for Inshaullah's Hejaz project, but there were some critical remarks as well. For instance, an article published in *al-Manar* had suggested that the donations collected from the Islamic world be spent for the education of Muslims instead of railway construction. *al-Malûmat*, on the other hand, had criticized the project on the grounds that it could instigate immigration into the holy lands.

When, in 1898, Inshaullah conveyed his thoughts to Selim Melhame Pasha in İstanbul through the Ottoman Trade Office in Karachi, the Pasha sent him a reply, which confirmed that his project had been received for the first time by Ottoman administrators.[117] The Pasha's declaration that he shared Inshaullah's view and that the issue was on the agenda in the *Komisyon-ı Ali* gave Inshaullah a great deal of hope.[118] The highly positive atmosphere triggered a new campaign in the press in favor of giving support for the Hejaz railway. "The *Observer* of Lahore wrote a leading article, and the *Paisa Ahbar* and the *Habl ul Matin* followed suit. Through the latter journal the idea spread throughout Iran, Russian Turkestan, and Afghanistan. From all quarters it was heard that the railway would be highly appreciated."[119]

Commencement of the Hejaz telegraph line installation in 1898 remotivated the Indian writer to bring up the Hejaz railway project, this time even more fervently. He got involved in polemical discussions in defense of the project; for instance, he heavily criticized a pessimistic article which had argued against the railway on the grounds that it had taken three years even to complete a telegrapgh line. Around the same time, the London-based *Times* published an article, which, quoting Gazi Osman Pasha and Münir Pasha, argued that Ottoman administrators were uncomfortable with German dominance over the Ottoman railway system[120] Encouraged in this way, Inshaullah wrote directly to the above-named Pashas and appealed to their Islamic patriotism.[121] Though no reply came to his letters, some promising developments had taken place in the meantime. For instance, in December 1989, *al-Malûmat* had published Ahmet Shakir Pasha's pro-Hejaz railway statement and its editor-in-chief, Mehmet Tahir Beg, had written a cordial and supportive letter to Inshaullah.[122] It is evident that thanks to his personal connections and writings, Muhammad Inshaullah had become an influential figure in his home country, India, in the Ottoman Empire, and in other Islamic countries.

Inshaullah, taking into consideration the financial crisis besetting the Ottoman Empire, suggested that the railway be financed by voluntary donations of all Muslims around the world.[123] We will discuss his significant contributions to this effort in the next section. As a matter of fact, it was partly thanks to his endeavours that the Hejaz railway project was discussed first at the Ottoman Marine Ministry and then the Council of Ministers. Looking at Izzat Pasha's report that has been referred to above, Inshaullah undoubtedly received his support to say the least.[124]

Abdülhamid attached great significance to all these proposals about "his long-standing dream" the Hejaz railway. Numerous military/political and religious factors which argued for the construction of the Hejaz railway had been put forth in those projects, albeit in a disorganized manner. They had also warned against possible dissenters, such as the British, the amirs of Mecca and the Bedouin tribes.

We have discussed the opinions of both insiders and outsiders who argued against the Hejaz railway project on the grounds that the British, the sharif of Mecca and the Bedouins would resist it. They also insisted that the Ottoman government had neither the necessary know-how nor the financial capability to run such a gigantic project. Now let us see how the Ottoman Empire mobilized funding for the Hejaz railway at a time when the Ottoman Treasury was in a severe financial straits.

CHAPTER 3

FINANCING THE HEJAZ RAILWAY

A) THE DONATION CAMPAIGN

It has been stated in the previous chapter that the Hejaz railway was perceived by foreign observers as a project that the Ottomans would not be able to accomplish. Given the overall cost of the railway— estimated to be 4 million Turkish liras, 15–19 per cent of the budget[1]—the diplomats commenting on this issue were by no means biased. At a time when no one knew how to provide kilometer guarantees for the Baghdad railway to which Abdülhamid attached great importance, how was this new and gigantic railway going to be financed? Budget deficits were such that even the military and the civil bureaucracy could not get their salaries regularly. Thus realization of the Hejaz railway project depended on finding untapped financial resources.

The Hejaz railway construction started immediately after *Ziraat Bankası* (the Agricultural Bank), a state bank, had released the 100,000 lira to cover the initial costs.[2] Most of this loan would be used to supply necessary materials required to start the construction.[3] Abdülhamid II took an important step by calling all Muslims to make donations to the railway construction. The caliph emphasized in his call that the Mecca line should not be seen as an Ottoman railway only; it would become the joint construct of the Islamic world. This is why the Hejaz railway had to be financed solely by Muslims and constructed by Muslim engineers. The power of Islam and solidarity among Muslims had to be shown to the entire world. In order to have no foreign control on this "sacred line", foreign capital had to be shunned. The caliph donated 50,000 liras, a considerable amount of money, for the promotion of the donation campaign. The newspapers of the period explained that the sultan's huge donation had mobilized the Islamic world and given momentum to the campaign.[4]

The main rationale behind the call for Muslims living outside the borders of the empire to contribute to the financing of the Hejaz railway was that this line would facilitate the hajj journey not only for Ottomans but for all Muslims.[5] Although published sources relating to the issue contain many generalizations, such as: "The Indian princes were competing to help. All the Muslims, from the poorest to those of modest means living on the Pacific islands accepted the caliph's invitation and invested significant sums of money", or "There is no country or village in Europe, Asia or Africa that would not be willing to help this holy duty by participating in the donation campaign."[6] Only 9 per cent of total donations were external.

The Donation Commission, established in order to run the donation campaign, systematically published certificates in various languages for donors. Moreover, prospective maps of the railway, in the form of cards and pamphlets, were printed and distributed for propaganda purposes.[7] Apart from these, donors would be honored with Hejaz railway medals. At the inauguration of the Tabuk line, 372 nickel *hatt-ı âlî* (sublime line) medallions were distributed.[8] There were gold, silver and nickel medallions. People who donated 5–50 liras were to be awarded with a nickel medal, 50–100 liras with a silver medal, and those who donated above 100 liras were to be given gold medals.[9]

Individuals who made extraordinary effort in raising donations were awarded a medallion of merit in addition to the Hejaz railway medal.[10] Some others who had contributed to the railway were honored with decorations and orders. Faik Pasha, the head of the

Beirut Delivery Commission of the Hejaz railway, and his staff were rewarded with a decoration and medallion for their outstanding service and effort.[11]

1- *External Donations*

Although the donations collected from the Islamic world outside the Ottoman Empire fell short of expectations, many Muslims from India, Egypt, Morocco and other countries had given close attention to the Hejaz railway. Donation campaigns in various Islamic countries commenced with the support of the local media and religious leaders. Local Islamic media, various religious orders, particularly the Nakhshibendi order, religious functionaries, landed gentry, merchants and other leading figures were among the ardent supporters of the donation committees. Ottoman diplomats in these countries also had an important function in the donation campaigns.[12]

a) *Donations Collected in India*

The most significant example of these Hejaz railway donation campaigns was the Indian one. The Muslims who carried out the campaign in India founded the Hejaz railway Central Committee with its headquarters in Hyderabad and offices in a number of cities. The committees founded in India began to collect donations right after the sultan's call. Muhammad Inshaullah was one of the leading figures in the press running these activities. As a matter of fact, Inshaullah claimed to be the mastermind of the Hejaz railway project. So much so that when the *Times* newspaper wrote that the founding father of the Hejaz railway project was Izzat Pasha, he reacted against this news by publishing a tract containing articles analyzing the emergence of the project. This tract tried to refute Lovat Fraser, the editor of *Times of India*, who had claimed that such a project could not have been launched by an Indian.[13]

Muhammad Inshaullah was the editor of the newspapers *al-Wakil* published in Amritsar and *al-Watan* published in Lahore. Inshaullah's newspapers not only advocates donations, they actually collected money. In 1904 he sent 1,050 rupees to Zihni Pasha, increasing the total amount to 17,600 rupees. Apparently he had sent money to İstanbul 13 times.[14] So, by February 1907, his donations amounted to 50,000 rupees and by 1908 they had reached 57,972 rupees.[15]

Apart from the donations that he mobilized and the propaganda he made via his newspapers, Muhammad Inshaullah himself was collecting donations. By 1910 he had collected 6,500 liras. We also know that Inshaullah was pivotal in delivering the Hejaz railway and merit medallions from the Ottoman Empire to their recipients in India.[16] Meanwhile, he had close ties with İstanbul and Arab media. He was so involved both in the domestic and foreign politics of the empire that he deemed himself authorized to express his views to a certain extent. For example he argued that the foreign policy of the Ottoman Empire had become too dependent on Germany and that this had to change. Inshaullah, who had reservations about Ottoman–German relations, suggested that Ottoman–British relations be improved.[17] He went as far as sending a letter to the Sublime Porte in 1909 threatening to withdraw Indian support unless the reasons for the mistreatment of Abdülhamid II who had just been dethroned by Young Turks was explained. Nevertheless, Inshaullah continued collecting donations under the rule of the Young Turks. As a matter of fact, another letter that he sent to the Sublime Porte read as follows: "I have sent the 36th installment as contribution to the Hejaz railway. The total amount of money that I have sent until now, including this installment, has reached 98,999 rupees 2 anes 5 paise, which equals approximately 791,993 kurush 80 para."[18] As a result, Inshaullah collected 3,243 liras by himself, which constituted 20 per cent of the donations collected in India.

Another important figure of the Hejaz railway Central Committee was the imam of Manarat Mosque in Mumbai, Abd al-Haqq al-Azhari. Azhari had learned of the Hejaz railway project from an Islamic newspaper called *Thamarat al Funun* published in Beirut. *Thamarat al Funun* was another newspaper that not only published pro-Hejaz railway articles, but also actually collected donations. When he read the news about the project, he was about to establish a printing house on the request of Mullah Abd al-Qayyum, a significant religious figure from Hyderabad, to publish Islamic works in Egypt. However, after being informed about the Hejaz railway donation, Azhari deemed it more appropriate to donate that money to the Hejaz railway and convinced Abd al-Qayyum to do so.[19] Azhari also intensively participated in the donation campaigns. Primarily he contacted the Ottoman consul in Mumbai and obtained a number of Hejaz railway certificates written in Urdu and Ottoman Turkish. Apart from that, he translated a letter by Abdülhamid II to Abd al-Qayyum into Urdu and made attempts to transfer some of the funds which were allocated in order to help the hajjis in Hyderabad to the Hejaz railway. While carrying on the donation campaign Azhari was also giving speeches in various meetings where he thought he could be influential. For instance, he made a speech in Amristar stating that it was high time for Indian Muslims to show their love for Allah and the prophet and make donations.[20]

When Azhari consulted Mullah Abd al-Qayyum on the issue of donating to the Hejaz railway instead of the printing house, Abd al-Qayyum welcomed this suggestion and said that he would do his best for the railway. Indeed he wrote letters to religious functionaries, notables, merchants and other prominent figures, established commissions and started collecting donations.[21] He managed to obtain permission from the Indian government. The first commission he established was in Mysore and he claimed to collect more than 50 liras a week. Later, he organized similar commissions in Madras and Malabar. Abd al-Qayyum continued his efforts in correspondence with Mehmed Emin Beg, the Ottoman consul in Mumbai.[22] And similarly, Mullah Abd al-Qayyum would also invite the Muslims of India to contribute to the railway via his speeches.[23] *Lisan al Hal*, a newspaper published in Beirut, described Abd al-Qayyum as the most prominent servant of the Hejaz railway.[24] Abd al-Qayyum collected precisely 1,736 liras on his own, an amount that constituted 13 per cent of the total. Consequently, it was those five names, that is to say, Haji Muhammad Mevludina, Muhammad Inshaullah, Gulam Muhammad, Abd al-Haqq al-Azhari and Mullah Abd al-Qayyum, who collected more than half of the total donations. The amount that Mevludina could collect was 1,360 liras.[25]

In India there were more than 150 Hejaz railway donation committees and nearly 100 of them were in Deccan. Apart from these, many people and organizations collected donations. For instance, the newspaper *Riad* in Lucknow collected more than 2,000 liras. One of the Indian princes granted 35,000 liras for the depot to be constructed in Medina.[26] The Ahmad Manshawi Pasha al-Kureishi Foundation in Karachi sent 2,000 Egyptian liras with a letter expressing their commitment and blessings to the Ottoman Empire.[27] There were significant individual donors as well. Bedreddin Tayyib, a prosperous man from Mumbai, was among those who struggled to collect donations.[28]

Some of the notable people of India set preconditions for the donation of large amounts. A Muslim prince from Hyderabad, for instance, donated more than 20,000 liras on the condition that the railway reach Medina.[29] The Sheik of Allahabad requested the next depot after Medina to be named Allahabad. If his request were fulfilled, he promised to donate 12,000 rupees to the Hejaz railway. According to *al-Watan*, the sultan accepted this demand.[30]

It was clear that Indians were interested primarily in the Mecca–Jeddah and Mecca–Medina lines. It was because the hajjis coming to Jeddah from the Hejaz would use only this section. The Mecca–Medina line was also important for them, for they wanted to visit the tomb of the prophet in Medina. Indian Muslims did not pay much attention to the lines that would pass from the north of Medina, that is to say between Damascus and Medina. This fact was highlighted from time to time in Indian media and put a dampner on the donation campaign. For example, an article published in the *Times of India* stated that if the sultan's intentions had been sincere, he would have taken into account the Muslim majority of India, Egypt, Morocco, China and Java that were to travel to the Red Sea by sea and thus give priority to the lines that would connect Yanbu and Jeddah ports to the holy cities. It was persistently emphasized that the more costly Damascus–Medina line was about to be finished and fellow Muslims from other countries would not benefit from that part of the railway at all. The fact that even the upper class hajjis from Yıldız Palace would have to travel to Jeddah by sea was used as an example to justify the priority of the Jeddah–Mecca line. There was an obvious warning that no new donations should be expected from Indian Muslims prior to the launch of construction to connect the Red Sea to Mecca.[31]

The exact amount of donations made by Indian Muslims is unknown. Estimates vary between 15,500 and 40,000 liras.[32] Failure to construct the Jeddah–Mecca and Mecca–Medina lines to which the Indians had attached great importance must have discouraged potential donors. Apart from this, it is not easy to estimate the extent of unfavorable British propaganda. The British media in India continually published articles against the Hejaz railway. It was claimed that money raised from poor Indian Muslims was being used to feather the nests of swindlers in Yıldız Palace instead of being used for the railway. Humiliating remarks such as, "The Hejaz railway is looked upon as a cash cow by the parasites at Yıldız Palace"[33] were rife.

The British attitude towards the Hejaz railway got more radical. Britain was concerned about every single initiative that could affect the passage to India and strengthen the Ottoman Empire. It was apparent that the Hejaz railway would pose a strategic problem to the British. The Austrian Ambassador stated that for the first time the British felt threatened in Egypt. Until the construction of the railway, Egypt was protected naturally by seas and deserts. But now it was only 250 km from Ismailia to Aqaba. Furthermore, if the Baghdad railway could be completed, Ottoman sovereignty would reach the Persian Gulf.[34] Indeed, with the incentive and support of Germany, it was written in the London media that the Baghdad and Hejaz railways were posing a threat to the British people in India and Egypt.[35] The British-imposed ban on wearing Hejaz railway medals was the most concrete example of British extremism. The British Foreign Ministry and the Indian Department forbade the Ottoman Ambassador Muzurus Pasha to award medals in India. Thereupon, Hejaz railway medals were sent to India surreptitiously.[36]

There were two main reasons why Muslims in this country made donations despite the negative propaganda circulating in the Indian/British media and all the British efforts in India. The first was that they regarded the Hejaz railway as the joint work of the Islamic world and believed that the Jeddah–Mecca–Medina line would be very useful on their way to hajj. The second reason was that the Hejaz railway was definitely an anti-Western project that reinforced the strategic position of the Ottoman Empire within the region and disrupted British plans in Arabia.[37] In other words, by contributing to the Hejaz railway, Indian Muslims had the chance to both demonstrate their traditional commitment to the caliph, thus creating religious solidarity, and express their discontent with the presence of the British in their country.

However, all these statements do not imply that the results of the donation campaigns in India were satisfactory. Many media outlets underlined this fact. *Naiyar-i Azam*, for instance, was very vocal: "The Indians are quite excited but the money raised is quite low, it's a shame, we have got such governments that can cover all the expenses on their own—especially Hyderabad—but the least amount comes from there." *Riaz ul Akhbar, Oudh Punch,* and *Nizam-ul-Mulk* published articles that criticized the apathy of the Indians and emphasized that a greater amount of money should have been raised from 70 million Muslims.[38] In fact, India ranked lower than Egypt in the fund-raising effort.

Although some of the newspapers in India such as *Oudh Akhbar* made negative propaganda by saying "we ourselves are starving, why should we help?" Muslim media played an important role in the campaign in general. Islamist newspapers in India were continually publishing articles concerning the legitimacy of the Ottoman caliphate and the donations. For example, the *Muhammedan* emphasized that every loyal Muslim was required to contribute to the Hejaz railway commensurate with their financial standing.[39] Abdülhamid had established close relations with the Islamist newspapers of India and had supported them in every way since 1880. The newspaper *Peyk-i Islam*, published in İstanbul by two Indian editors who were exiled by the British government in India because of their anti-British activities, can be cited as an interesting example. From the first issue of *Peyk-i Islam*, the British embassy observed that the newspaper aimed at developing Islamic unity against British rule.

London also referred to the content of the issue as being "incompatible with friendship" and the issue was discussed in Parliament. Following the British reaction, *Peyk-i Islam* had to cease publishing.[40] Another feature of the newspaper was that it was published by a Muslim from Punjab, to be read in India, in the private printing house of Abdülhamid at Yıldız Palace.[41] In short, Abdülhamid's Islam-centered relations with the Muslims of the country played an important role in the Indian Muslims' interest in the Hejaz railway. The Indian Muslims' commitment to the Ottoman caliphate was already apparent. An example of this would be the financial and spiritual support the Indian Muslims had given to the Ottoman Empire during the 1877–78 Russian War. This support continued even after Abdülhamid was deposed. In a letter sent by the Young Muslim Community of Bengal, India, the establishment of an Ottoman Center of Commerce to display Ottoman industrial goods was requested—a suggestion that was welcomed by the Ottoman Chamber of Commerce. The letter also stated that the donations to the Hejaz railway would continue.[42]

b) Donations Raised in Egypt and Other Islamic Countries
The fact that the Egyptians showed their commitment to the caliph by donating to the Hejaz railway disturbed the British occupation forces more than the Indians' activities, as Egypt was still legally a part of the Ottoman Empire. Therefore pro-British media in Egypt commenced an intense anti-Hejaz railway campaign under the guidance of the British. Claims that Egyptians were making sacrifices for an unrealistic project and that, just as in India, the donations would be used for Abdülhamid's personal expenses, were also asserted in Egypt. The British did not ban contributions to the railway project outright, for fear of the reaction that might have occurred, but by the command of Lord Cromer, they at least prevented officials from joining the donation campaign.[43] They also refused the Ottoman government's request for a 5 kurush donation from every Egyptian. This responsibility was, however, imposed on every Muslim family considered to be subjects of the Ottoman nation.[44]

The Hejaz railway donation committees, which were established the same way as in India, sent donations to İstanbul via the Egyptian Extraordinary Commissary.[45] One of the most successful committees that had been established throughout Egypt was the one set up in Harbiye. Ahmad Pasha al-Manshawi, who was the chairman of the Harbiye committee, later became the chairman of the central committee that collected donations for the Hejaz railway. The amount of donations that Ahmad Pasha made during this process was over 2,500 liras.[46]

With Khedive Abbas II at their head, Egypt's high state officials, religious functionaries and orders—especially the Naqshbandi order—merchants and the rich, and such middle class constituents as engineers and doctors all supported the donation campaign. Khedive Abbas promised to grant some part of the equipment needed in the construction of the Hejaz railway. The fellahs who belonged to the low-income group were making contributions as best they could.[47] The same factors that motivated the Muslims in India were also effective among the Muslims of Egypt.

The donation campaign was supported passionately by Egyptian newspapers. For instance al-Siba, a local newspaper in Egypt, not only supported the Hejaz railway donation campaign through its writings but also started its own donation campaign. Nearly all the newspapers published in Cairo supported the campaign. The amount of donations raised by al-Muayyad was over 1,000 liras. Mustafa Kamil, the editor of al-Liva, was rewarded by Abdülhamid for his achievement in raising donations.[48] Articles that sincerely supported the campaign were issued in newspapers such as al- Raid al- Misri, al-Ahram, and al-Manar.[49] Another interesting practice in the Hejaz railway campaign was that the names of the donors and the amount they donated were made public.[50] However, despite all the efforts stated above, the amount of money collected in Egypt was not sufficient. Still, it is very remarkable that the donations raised in Egypt were a third more than those raised in India, where the population was seven to eight times bigger than that of Egypt.[51]

The chief engineer Meissner says that at the end of 1900 external donations had reached 13 million francs.[52] Apart from India and

Egypt, funds were collected in other Muslim-populated countries such as Morocco, Bukhara, Russia, China, Dutch India (Java, Sumatra), British Guiana and Trinidad.[53] The 8,768 liras from the Moroccan Amir Abdul Aziz had set an example for other administrators. Officials like Debbag Efendi from Mecca were sent to Morocco by the Ottoman Empire to collect donations.[54] Kologlu states that the donations from Morocco and Bukhara depended solely on the decisions of the sovereigns. Morocco donated 8,800 liras, and Bukhara, 22,500 liras.[55] In Russia, Muslims in the Crimean region in particular contributed to the Hejaz railway. The Russian consul Zinoviev's declaration that he had sent 160,000 rubles via the consuls of the Bukhara Amirs indicated that Russia acted in a non-obstructive way from time to time.[56] However, it is understood that although the Russian government had not publicly prohibited the donations, administrators were expected to discourage them. After learning of the Russian administration's approach to the Hejaz railway campaign, the British ambassador O'Conor suggested that the British conduct such a practice in India.[57] The French government's approach to the campaigns in its colonies was harsher than that of the British and the Russian governments. France would give a limited number of hajj permissions to Algerian and Tunisian Muslims. The French consul of Jeddah wrote in a report he sent to the Minister of Foreign Affairs that they had to prevent as many of their Muslim peoples as possible from traveling to hajj by imposing health and financial obstacles.[58]

The Ottoman government used various means to keep the Islamic world's attention on the Hejaz railway and maintain the flow of donations from outside the Ottoman Empire. The British kept up their anti-Hejaz railway campaign. The British media in Egypt, for instance, was writing that the Ottoman government could not carry on railway construction because of scarcity of resources.[59] The assertions that the construction of the Hejaz railway was only a dream and that money raised for the project was being leaked to the "Yıldız parasites" had to be refuted. Even the slightest corruption could lead to a scandal in the world of Islam. For this reason, the public was informed of each and every phase of the Hejaz railway construction through weekly or monthly official announcements.[60] In spite of all the provisions, we should state that allegations of corruption could not be avoided. The *Times of India*, for example, referred to Izzat Pasha and his immediate acquaintances as the "Yıldız group" and claimed that this group had made a fortune out of the Hejaz railway.[61] The issue was mentioned in an Ottoman report drawn up by the British government in 1906, where it was depicted in a rather interesting way: "The Bulk of the enormous fortune that he must undoubtedly have amassed is safely invested in Europe, and every possible preparation made for a rapid flight in case of necessity."[62]

Foreign observers also described Hasan Pasha as a chronic kleptomaniac who filled his pockets with the government's money.[63] Hasan Pasha was accused by some Ottoman administrators, as well. For instance, Hejaz Governor, Ahmet Ratib Pasha, did not hesitate to convey to Abdul Hamid that if Minister of Navy Hasan Pasha, who regularly got a 20 per cent commission fee out of each tender, continued to be the chief member of *Komisyon-ı Ali*, no one would give any donations to the railway.[64]

2- *Donations Collected in the Ottoman Empire*
a) People and Institutions that Contributed to the Donation Campaign
The donation campaign within the empire was organized through commissions under the chairmanship of the regional governors (*kaimakams*).[65] Abdülhamid II opened the campaign by making a 50,000 lira donation; his statement that he would be very pleased to see contributions by state officials stimulated senior officials. At the Meclis-i Vukela it was decided that all officeholders, inspired by the sultan, would donate one month's salary to the Hejaz railway.[66] Although the decision sounded like a suggestion, within the political atmosphere of those times, none of the senior officials were expected to challenge this decision. Indeed, it is understood from the documents of the year that the officials, following the sultan, had agreed to give their "one month salary as donation very enthusiastically", and that this was accepted as a fact.[67]

Apart from the senior government officials, many military officers, notables, religious functionaries and merchants supported the campaign.[68] Every rank within the Ottoman military donated to the Hejaz railway. Ahmet Muhtar Pasha, the Extraordinary Commissary of Egypt, was rewarded with a gold medal for his donation.[69] General Abdullah Pasha had initially donated 9,000 liras and later made an 1,800 kurush donation; thus his silver medal was replaced with a gold one.[70]

The governors of some of the provinces had promised to help the Hejaz railway long before the donation campaigns started. The governors of Beirut, Aleppo and Syria, for instance, had each promised to donate 40,000 lira. Syria had pledged to donate an amount that equaled one-tenth of the province's annual tax. Bursa had promised to send 75,000 liras. But the biggest commitment was made by the province of Hejaz, which had promised to send 200,000 liras.[71] Besides the provinces, *kazas* (districts) also gave commitments to raise monies. Kırcali kaza, for instance, promised to donate 40,000 liras, Hamidiye 20,000, and Köysancak 5,000 liras.[72] Although the governor's offices started to send the donations collected within their boundaries, it would soon be seen that the contributions were not near the amount that was promised in the beginning. The total sum that came from the Hejaz province was an example of the fact that that the amounts actually sent were much less than expected.[73] The protocol of Meclis-i Vükela complained that only 350,000 liras of the promised 800,000 liras could be raised and emphasized that provinces had to be warned about their initial commitments.[74]

In addition to the provinces, ministries and other government agencies had promised to donate to the Hejaz railway. The financial backing for the government agencies' promises was the salaries of the personnel working there. Some of the officials working in various agencies preferred to make individual contributions to the Hejaz railway. Mayor Abd al-Qadir Qabbani donated 30 liras to the Hejaz railway. Qabbani was an editorial writer for the newspaper *Thamarat al-Funun*, which was mentioned above in terms of its role in the Hejaz railway donation campaign.[75]

The Meclis-i Vükela had decided to issue and distribute two million liras worth of donation tickets for civilians outside the government agencies and government personnel to be able to make donations.[76] If anyone desired to make an "abundant" donation they would be given a receipt by the special commission, the *Komisyon-ı Mahsusa*.[77] It was natural that all large donations to the Hejaz railway were made by higher income groups. For example, Abbud and Halbuni, two merchants from Beirut, decided to donate 2,000 liras, a very substantial sum. However, they agreed to pay at the end of every 100 km of construction. Sultan Abdülhamid praised their generosity. This method of paying in increments as the railway progressed was also adopted by Ali Beg al-Katib, who was a merchant in İstanbul, and Mahmud Pasha al-Jazairi, one of the notables of Damascus.[78]

There were also people who donated materials instead of money. A number of cities, like Salonika and Aydın, had granted a substantial amount of traverse. While people from Menteşe donated 7,000 timbers, Eregli coal enterprises donated 458 tons of coal. There were also donations made by the Christian subjects. The bishop of the Orthodox church in Beirut called upon his community to donate to the Hejaz railway. Articles in favor of the Hejaz railway construction were issued in the *Lubnan* newspaper, which was published by Christians. The French DHP Railway Company, Hejaz railway's rival in the Syrian region, donated 84 liras to the Beirut donation commission.[79] The Armenian İzmit delegation reported a 3,000 kurush donation.[80] "Lord Stanley who was a member of the House of Lords and who was known for his amicable feelings towards the sultanate had also donated 1,000 liras via the embassy in London."[81]

In order to show the significance of the donations—or in other words, in order to be able to compare the level of sacrifice—it would be useful to give examples of the annual incomes of some parts of society. For instance, the annual income of a day-laborer in Syria was 10 liras and that of a qualified worker 20–30 liras. The annual income of foreign qualified workers could be up to 60 liras. If we were to give examples from the Ottoman military, the total annual income of a lieutenant was 50 liras and that of the brigadier was around 600 liras.[82] In this regard, when the average income level of the society is considered, it is clear that even a 10 liras donation asked for a significant level of self-sacrifice. We see that for a day-laborer to donate 10 liras, he had to give up his annual income.

There are claims that from time to time the monies collected were far from being voluntary donations but were almost made obligatory. Some French observers, for instance, testify that in some regions of Syria donations were collected by force. According to Slemman, who was one of these observers, the minimum amount that every donation commission of a district in Syria had to collect was pre-determined. In order to achieve the determined amounts, the commissions would set the donation share of each village arbitrarily. Furthermore, the religious affiliation of the inhabitants, Muslim or not, was not taken into consideration when these settings were made. If need be, the amounts in question were collected with the help of the military police and sanctions such as detention were implemented in case anyone tried to avoid "donating".[83] On the certificates given to the donors there was a statement saying that the owner of the certificate had made a donation, and a second donation would not be asked of him. This fact was cited as an indirect proof of forced donation in Beirut. In need of urgent donations, the governor of Beirut would command the tax records to be checked to see whether the wealthy people had made donations to the Hejaz railway or not.[84] Again, in some documents sent to the Grand Vezirate it was stated that some people had avoided buying the donation tickets. The question of how these people should be treated was also asked in these writings.[85]

Although it was not necessarily a forced donation, we see that the ones who promised to donate a certain amount of money could not run away from it. The case of one Ali Efendi, "who had left his office without paying the 720 kurush which he had promised to donate, and was later found to have been promoted to the district governorship of Sungurlu, the district of Ankara. A written notice of the Ministry of Internal Affairs then asked for the collection of the aforementioned amount" is an interesting example in this regard.[86]

For those who had promised to donate one month's salary but failed to do so, it was decided to cut 5 per cent of the salaries each month in order to collect the promised sum.[87] As for those who had passed away before donating what they had promised, it was decided that it would be inappropriate to put pressure on their survivors. However, appealing to their religious feelings was deemed suitable.[88]

b) The Organization of the Donation Campaign

The Donation Commission in İstanbul had undertaken the organization of the donation campaign. The Commission was established under the presidency of the Minister of Finance and all the activities were performed within the Ministry of Finance.[89] The issuing of donation tickets, recording of donations from both inside and outside the country, etc. were all duties of the Donation Commission. In 1904 it was decided to organize an administration named the "Hamidiye-Hejaz Railway Fiscal Administration" in place of the Donation Commission. This decision was justified by the necessity of keeping a different account for every source of income—a task that the personnel working on the Donation Commission could not spare the time to do because of their workload.[90]

The Central Donation Commission assigned the fundraising activities to provincial commissions. The provincial commissions consisted of mayors and local notables under the supervision of the region's highest administrative authority.[91] The provincial donation commissions acted as mediators between the central government and the local Muslim communities. The fact that the government envoys and local gentry came together in these commissions and were totally committed to their duties increased the level of success in collecting donations.[92] The composition of the provincial donation commission in Beirut was well-known. An examination of the professions of the commission members would be useful in giving an idea of the character of the commissions. The Beirut Commission consisted of Yahya Bey, the commissioner of the gendarmerie; Muhammad Efendi Mustafa Beyhum, a distinguished notable; Khalil al-Hisami, the chief clerk of the city council; and Khalil al-Barbir, employed at the Port of Beirut. Supposedly, the practices in the other provinces developed in a similar way.[93]

The provincial donation commissions were in close collaboration with the village headmen (mukhtar) and imams. This enabled the donation campaign to reach even the remotest districts and villages.

In this way it became much easier to explain to people why Muslims should contribute to the railway. Another benefit of the mukhtars' and imams' involvement was that they could make a list of the names of those wealthy enough to contribute to the railway and then send this list to the related commissions. When the money they collected reached a certain amount it was sent to İstanbul via the Ottoman Bank or the postal service.[94]

The central and provincial donation commissions were kept under tight control. A board of three men assigned by the sultan was responsible for supervising the Central Donation Commission. The Supervisory Board reviewed the provincial commissions through justice and finance audits. The number of violations remained low throughout the construction process, which indicates that the government was highly meticulous on the surveillance issue. Besides, most of the audits concerned the use of donated money in areas other than railway construction. For example, 59,019 liras of the money raised in Trabzon was transferred to the head of the provincial treasury (*defterdar*), Abdurrahman Şevki Efendi, who used the money elsewhere.[95] Mustafa Beg, who was a member of the gentry, was put on trial for charges of bribery.[96]

c) The Role of the Media in the Donation Campaign

It should be emphasized that the role of the Ottoman media was very important in the success of the Hejaz railway donation campaign conducted within the borders of the Ottoman Empire. When the limited means of communication is considered, it is understood that the most important media to explain to people the objectives of the Hejaz railway were the newspapers. The newspapers *İkdam* and *Sabah* actively supported the Ottoman government on this issue. Both of these newspapers were continually writing about the benefits that the Hejaz railway would bring to the world of Islam. How the Hejaz railway would facilitate hajj, one of the five pillars of Islam, and thus multiply the number of people going to hajj were topics that were continually written about in these newspapers. Apart from the railway's religious dimension, the newspaper would also write about the economic and social benefits that the railway would bring to the

Ottoman Empire in an exaggerated way. It was as if a new period would begin, all the mines would be processed, new industrial plants would be established and the power of the Ottomans would once again be revealed. Furthermore, the Bedouins who lived in the holy land where nomadic life was common would be able to adopt a sedentary life by means of education, and "every depot of this sacred line would function as a cultural center".[97] We know that articles based on similar arguments were issued in local newspapers like *Thamarat al-Funun*.

Dozens of articles that appeared in the the newspaper *Tercüman-ı Hakikat* between April 2 and August 11, 1904, can be shown as a concrete example in this regard. A series composed of 21 articles by Muktesid Musa (Economist Musa) was published in the newspaper. The title of the series was, "The Great Hejaz Railway's Spiritual and Financial Meaning". According to the author it was impossible to claim a greater work in the world of Islam than the Hejaz line. The Nile channels which were constructed 5,000 years ago, the pyramids, Semiramis waterways, the hanging gardens of Babylon, the Baghdad pavilions, the Al-Hamra castle in Spain, the roads and works constructed in Eastern Europe could all be counted as negligible when compared to the Hamidiye Hejaz line. What was the benefit of a pyramid? The channels would only be beneficial to the related villages. Muktesid Musa wrote in his article issued on August 11, "Donations are necessary for the construction of the Hejaz railway to finish. Everyone should help in an initiative that will bring happiness to three hundred million people." The author also described the Hejaz railway as "our treasure that deserves to be recorded in Ottoman history with precious stones" in an article published on July 7.[98]

Apart from the newspapers, there were also books published in both Turkish and Arabic that talked about the benefits the Hejaz railway would bring and that called the people to make donations. The books of Mehmed Şakir and Mehmed Emin, namely *Beşâirü'n-Nehhâz min Hattı'l Hicâz* for instance, is a good example of works written in Arabic that propagandized for the Hejaz railway. A closer examination of

Muhammad Arif's manuscript, titled *The Increasing and Eternal Happiness – the Hejaz Railway*, can shed light on the issue.[99]

Muhammad Arif was one of the Arab writers who had devoted himself to Abdülhamid's government. He was in close relations with the pro-Ottoman Arab governors both in Damascus and Istanbul. According to his writings he decided to write a book about the Hejaz railway when he visited Istanbul in August of 1900, in other words two months after the sultan's will had been publicly announced. The Ottoman administrators in Istanbul must have played an important role in the author's writing process. This can be said because in the book there is detailed information on the issue, thus leaving no place for doubt that he had been officially assisted in the process.

As an author from Damascus who was in favor of the Ottoman government, it was natural for Muhammad Arif to be interested in a project that would connect Damascus with the holy lands. The fact that the manuscript was finished in a hurry within the year 1900 makes it clear that the main object of its being written was to propagandize for the Hejaz railway. We know that this was a time when Arab nationalism in Syria and the Hejaz was being born and developing with English support. In Hejaz railway propaganda, Islam was emphasized against Arab nationalism. The Hejaz railway donation campaign had been launched and the campaign struggled to receive financial and spiritual backing in the media. In other terms, Arif had once again chosen to be on the Ottoman government's side. Arif's manuscript took its place in the library of Yıldız Palace as soon as it was written, which also indicates that it was part of the official Hejaz railway propaganda.

It is easy to understand that the main target audience that Muhammad Arif wanted to affect through his writings was the Arabs who lived in areas through which the railway would pass. The objective to refute the arguments which were a part of the propaganda conducted against Ottoman sovereignty in the region and which were against the Hejaz railway stands out in Arif's work. In his 157-page work, Arif lengthily explained the benefits of the Hejaz railway. While he mentioned how the railway would facilitate the hajj journey and its motives to serve the holy land, he also meticulously examined the contributions that this railway would make to the development of agriculture and industry, the increase in population and economic stability in Arabia. In short, Arif was trying to convince the inhabitants of Syria and the Hejaz that the Hejaz railway was a highly beneficial project for them. After successfully giving reasons why people should support the railway, he would then give the sultan's 50,000 lira donation as an example and emphasize that every Muslim had to contribute financially to the Hejaz railway. The donation call was directed particularly at the Muslims of the Hejaz. This was because, according to Arif, the Hejaz railway would provide great advantages to everyone from every social rank, especially the peasants and the Bedouins. Again thanks to the Hejaz railway, mining and agriculture would develop and the people of the region would find themselves living better that they had ever dreamed possible. The people of Mecca and Medina would have the chance to sell their fruit in distant markets without making an effort and for the ones who lived near the railway there would be more business opportunities.

While explaining in his work the benefits of the Hejaz railway, Arif put excessive emphasis on the contributions the railway would make to the Bedouin tribes. Arif wrote this long chapter in detail and paid great attention to sound convincing. Arif's sensitivity stemmed from the thought that the greatest danger for the Hejaz railway would come from the Bedouins. Indeed, the main reason that lay behind the Bedouins' rejection of the railway was the fear that they would lose the money which they earned from hiring camels to hajis or by plundering the hajj convoys. They also feared that the money that the sultan sent under the name of *urban surra* to deter them from plundering the hajjis would stop. Arif was striving to refute the Bedouins' fears that they would lose revenue with detailed calculations and by giving numbers. Firstly, owing to the railway, the Bedouins would have the opportunity to market their animal skins and wool in far places and also at a higher price. At the same time they would be able to generate an income by selling their dairy products at the depots. Arif agreed that after transportation through the Hejaz railway started hajjis would rent camels only for short distances. However, the number of people would be much higher than before, so the profit gained from renting camels would not fall as feared, but would even rise. In addition, the Bedouins could also earn their incomes by working in services related to railway management.

Muhammad Arif knew that the Hejaz railway was being constructed in order to reinforce Ottoman sovereignty over the region and that the Bedouins were aware of this fact. He tried to explain to the Bedouins that a stronger Ottoman sovereignty would not be to their detriment. On the contrary, it would work against the common enemies of the Muslims. The reason why Arif spent so much of his time and energy on convincing the Bedouins would become much clearer in the coming years, when the Hejaz railway could not be completed because of Bedouin resistance. As a matter of fact, in 1913 the Arab thinker Shakib Arslan, another Ottoman supporter, wrote to Talat Pasha that he had met with various Bedouin leaders and was trying to convince them on the issue of extending the Hejaz railway to Mecca.[100]

Another thing that Muhammad Arif complained about was that more and more hajjis had started to go to Jeddah by sea rather than using the pack trail that started from Syria. This fact naturally reduced the income resources of the region and upset Arif, as he was a man from Damascus. Hajj had for generations been the mainstay of Syria, commercially and financially. Arif was writing that he would be very glad to see the hajj's contribution to the region rise with the increasing number of hajjis due to the Hejaz railway. Clearly Arif's work is a concrete and successful example of Hejaz railway propaganda aimed at Arabs.

B) OTHER INCOMES

Although the amount of internal and external donations was considerable, the continuation of the construction depended on finding new income sources. Apart from voluntary contributions, the state was to confiscate one month's salary of every government official—in effect cutting small amounts from the officials' salaries each month. Non-muslims who were included at first were exempted in 1902.[101] Also a decree issued in 1903 stipulated that the May salary of all officials with income over 500 liras was to be cut by 10 per cent annually. In addition, officials' March salaries would be cut in the name of Hejaz railway donation.[102] Moreover, a 10 per cent cut was applied every March to the salaries of officials with salaries over 2,500 liras in the name of donation for Harameyn inhabitants. In 1903 this amount was 640,690 kurush and, after distributing to the public, the remaining amount was transferred to the Hejaz railway as usual.[103]

Apart from the Ziraat Bankası loans, donations and cuts from officials' salaries, a new tax was created in order to ensure regular revenue flow to the railway. The head of every Muslim family was a taxpayer and was supposed to pay 5 kurush annually. Another interesting income source was the skins of lambs slaughtered during the Feast of Sacrifice. In 1901, Muslims were asked to donate the skins of their slaughtered animals via commissions in their regions.[104] Given that between 80,000 and 150,000 animals were sacrificed each year,[105] the governors projected a high yield. In fact, 40,000 liras was collected this way in 1901.

Another important decision made it obligatory for all official documents to bear donation stamps. Because the issuance of donation stamps was delayed for a while, the relevant government official wrote "donation has been taken for the Hejaz railway" in the back of the document and ratified it with his personal seal.[106] However, foreign companies operating in the Ottoman Empire protested against this. The Lynch Company, which engaged in ship management on the Tigris and Euphrates rivers, did not want to use Hejaz railway donation stamps in its operations. Officials working in the Austrian and Italian consulates expressed similar concerns. Notwithstanding the foreign companies' resistance, they could not secure an exemption for some time.[107] Later, diplomatic pressure bore fruit and the stamp obligation was bent in favor of foreign companies. As a matter of fact, in 1903 it became obligatory for Ottoman subjects but voluntary for foreigners to pay 2 kurush for their passports and custom house seals to support the Hejaz railway.[108] All requests sent via the consulates were completely exempted from this tax.[109] Interestingly, informers were asked to pay 40 liras for each denunciation. However those which were considered of vital importance "for the interest of the state" were excluded.[110]

The right to issue and distribute all the official documents used in the empire was transferred to the Hejaz Railway Administration. Thus various valuable documents were subjected to Hejaz railway stamp duties.[111] In addition, 15,000 liras deducted from prescriptions written by Dersaadet pharmacists was to be contributed to the Hamidiye Hejaz railway expenses.[112] Moreover, a 1 kurush Hejaz line stamp had to be affixed to any published book.[113]

Starting from the 1900–01 hajj, a certain amount of contribution was raised from hajjis who arrived at Jeddah, unless they were extremely impoverished. Afterwards hajjis were obligated to pay a certain fee in return for using the Syrian and Egyptian roads. Meanwhile, customs duties collected from the Port of Beirut were reserved for the Hejaz railway. Another source of income was the profit made through the transactions of bringing Riyals that would soon be out of circulation in Lebanon and Yemen and sending Ottoman Mecidiye in return.[114] Besides, jewelry bearing the sultan's signature (tugra) which varied in size and value was produced.[115] The money thus raised would be transferred to the Hejaz railway by the Currency Administration (Meskukat İdaresi). The Currency Administration was also produced beautifully designed arms and sold them for 5, 10 and 20 kurush.[116] The expenses of the railway kept rising, so in 1905 it was decided to sell certificates to students studying in formal schools. In keeping with the decision, primary education certificates would be sold for 5 kurush, secondary education diplomas for 10 kurush and higher education diplomas for 20 kurush and this would create another source of income.[117]

The Ottoman government developed another method of obtaining income by selling stamps of historical value at auctions through their foreign missions.[118] Old artillery, iron implements etc. from the historical castles of Acre, Jaffa, Sayda, Sur and Tripoli would also be auctioned.[119] Auctioning of objects taken from shipwrecks in the Mediterranean and the Black Sea was another interesting fundraising method.[120] Likewise, postcards used in public post offices, stamps and wallets with photographs of a locomotive on them, matchboxes and maps would be put up for auction. In 1908 the Ottoman diplomat Monsieur Kemrih declared that he would pay 1,600 liras annually if he was authotized to put a label "in support of the Hamidiye Hejaz railway" on matchboxes for 15 years. His suggestion was accepted by the Council of Ministers.[121]

When the government found out that the maps that were sold to contribute to the Hejaz railway were being counterfeited, they stated that no one would be allowed to take advantage of the monies allocated to such a religious and auspicious work. Thus the government adopted a preventive action against fraud.[122] In addition, the official authorities offered to organize a lottery draw with a prize of 120,000 liras to support the "construction of the Hejaz railway".[123] The foreign companies' objection to the use of donation stamps on their theatre tickets was accepted but local companies kept paying.[124] Another measure taken in order to cover the deficit of over 200,000 liras in 1907 was to sell the Road Administration in Beirut.[125] Half of the annual forest tax had also been allocated for the Hejaz railway.[126] A significant portion of the telegraph line revenues would be earmarked as well.[127] Lastly, donations were also raised via cigarette papers.[128]

Abdülhamid equipped the Komisyon-ı Ali with several important privileges hoping that in the long run it would become a source of revenue. On the 20 km area on both sides of the Hejaz railway the concession to extract and operate mines belonged to the Hejaz railway, similar to the Es Salt phosphate deposit and Hamam sulphur deposit.[129] The electricity generation revenues from Yarmuk and Jordan were reserved for the Hejaz railway. Again, the concession to operate mines on the Jordan Plain on the coasts of the Dead Sea was granted to the Hejaz railway. The government expected to secure an income of 3–4 million francs by selling some of the mines along the river that could be used in industry and medicine.[130]

Apart from the direct sources, funds from some of the concessions and tenders were mobilized for the Hejaz railway as well, such as half of the İnegöl mineral water revenues. Similarly Manuel Salem Efendi was permitted to produce and sell cigarettes provided that he made an annual contribution of 3,000 liras to the project. The concession agreements of those who had been reluctant to contribute to the Hejaz railway were repealed. For instance, when a company which had held the electricity and tramway concessions in Beirut refused to give 10 per cent of its revenue to the Hejaz railway, the sultan refused to sign the agreement.[131] There were also times when concessions were withheld for various reasons. For instance, the concession to rent ferries granted to the İzmir Hamidiye Ferry Company was cancelled on the grounds that the company was importing forbidden arms and harmful documents.[132]

The forced labor service liabilities for road construction in Nablus and Acre was liquidated and the revenue transferred to the Hejaz railway. Immediate cash requirements would be covered by loans released by Ziraat Bankası. For example, on March 28, 1904, when the salaries of workers could not be paid, Ziraat Bankası was called for funding.[133] In the meantime the donations collected for the Hejaz railway were deposited in the Ottoman Bank with an interest rate of 2.5 per cent, to generate extra income.[134]

C) FINANCIAL SOURCES EVALUATED

The operating income obtained from active lines naturally contributed to the financing of the railway. However, until 1909 the total operating income was a mere 225,310 liras[135] and this constituted only 5.7 per cent of the 3,975,443 lira total revenue of the Hejaz railway. Until that year, a 485,993 lira loan was borrowed from Ziraat Bankası, which covered 12.2 per cent of the funding. The lion's share—1,648,692 liras, which accounted for 41.5 per cent of the total resources—was collected through taxes and mandatory payments.[136]

Donations to the Hejaz railway—which were 1,127,894 liras in total—were the second biggest source. This constituted 28.4 per cent of the total.[137] If the revenue obtained through the sale of animal skins is added, the ratio reaches nearly to 34 per cent. Apparently, Muslims from all around the world financed one-third of the railway, dubbed the "sacred line". The Austrian Ambassador Pallavicini described this as "an unprecedented fund raising method in the modern economy".[138] Reading Abdülhamid's statements below, the amount of donations even exceeded his expectations. Referring to Izzat Pasha, Abdülhamid said, "I've been amazed by how he managed to collect the necessary money for the Hejaz railway from all Muslims around the world, and especially the Indians, in such a short time".[139] However, we have to reiterate that Abdülhamid's approach was far too optimistic when it came to donations from outside the empire. The amount of external donations was 110,000 liras and this constituted only 9.5 per cent of total donations and a mere 2.8 per cent of total funding.[140] So this interesting dimension of the Hejaz railway's financing should not be overstated. For if we consider the Ziraat Bankası loans to be a form of tax, as its capital is derived from taxes, more than half of the Hejaz railway funding can be said to come from obligatory payments by Ottoman society.

The donations which constituted 28.4 per cent of the total funding in 1909 had previously made up the biggest share. In 1903 donations represented 63 per cent of the total, in 1904 it was 55 per cent, in 1905, 35 per cent, in 1906 again it was 35 per cent and finally, in 1907, it was 32 per cent. This was followed by a parallel decline in the amount of donations. From 1900, the year when Abdülhamid had launched the donation campaign, until the end of 1901, 417,000 liras were collected. In the following two years only 329,000 liras were raised. Within the period of 1904–05, the amount of donations declined rapidly and only 74,000 liras could be collected. Consequently, a Meclis-i Vukela resolution of March 11, 1906 urged that the fundraising process be sped up. During the second biggest donation campaign that started in 1906, 308,000 liras was collected in three years.[141] Obviously, the exciting momentum of the earlier years could not be sustained. After 1908 there were no major donations. After the completion of the Medina line, resources consisted mostly of operating incomes, taxes and related obligatory payments and funds granted to the Hejaz railway.

Clearly, Abdülhamid tried interesting methods to generate alternative resources to complete the Hejaz railway. The fact that the line reached Medina in 1908 proves the effectiveness of his methods.

When the average rail length constructed annually is taken into consideration—which was 121 km on the İzmit–Ankara railway, 73 km on the Salonika–Manastir line and 100 km on the Alasehir–Afyon line—it can be said that, at 150 km a year, the Hejaz railway was constructed at above-average speed. Alternative fundraising methods and a minimal kilometer cost played an important role in this success story.[142] The kilometer cost of the Hejaz line was approximately 35,000 francs[143]—one-third of the average. The land structure also played an important role. For instance, between Damascus and Maan there was only a 120-meter tunnel and a great viaduct in Amman. The Haifa branch line, which required a considerable number of tunnels, bridges and viaducts, increased the cost to 100,000 francs per kilometer. This example shows the advantages the Damascus–Medina main line brought.[144] However, the main reason behind the low construction costs was recruitment of Ottoman soldiers as labor force: because soldiers' remunerations were paid by the Ministry of War, the Hejaz railway budget was relieved of this burden. The fact that both the expropriation process and the supply of stone and wood for the construction of bridges were cost-free also contributed to the savings.

CHAPTER 4

THE CONSTRUCTION OF THE HEJAZ RAILWAY

A) THE HEJAZ RAILWAY ADMINISTRATION

The continuation of the railway construction in a place quite far from the center of the empire depended primarily on effective organization. The main decisions on construction work had to be made in İstanbul but it was also necessary to get established in Syria and Hejaz provinces to manage the construction operation and align the activities of the two.[1]

In 1896 Abdülhamid had appointed Nazım Pasha, the governor of Syria, to the chairmanship of the local commission established in Damascus, the starting point of the railway. Nazım Pasha had held office until he resigned in 1908. Another important member of the commission was Marshal Kazım Pasha, the commander of the 5th Army in Syria. In late 1901, Kazım Pasha was appointed as the head of the Hejaz Railway Construction Ministry. Sadiq Pasha al-Muayyad, a military attaché of the sultan, was appointed as Kazım Pasha's chief assistant. Sadiq Pasha had just fulfilled his duty as manager of the telegraph line installed along the Hejaz railway. Therefore, he knew the region's topography very well.[2]

The remaining members of the Damascus Commission were: Adib Nazmi Beg, who was secretary of the city council and the editor of the magazine *Syria*, Amirü'l-hacc Abd al-Rahman Pasha, Marshal

Cemal Pasha and Muhammad Fawzi Pasha al-Azm who belonged to one of the most prominent families in the city.[3] The commission was composed of senior government officials of the region and representatives of prominent Damascus families. In this way the government intended to gain the support of the local powers in the construction process.

The Damascus Commission responsible for the construction was being directed and controlled by the *Komisyon-ı Âli* (Supreme Commission) in İstanbul. The most important ministers and administrators of the Ottoman Empire had worked in this Commission. For instance, the Minister of Public Works Zihni Pasha, the Minister of Finance Tevfik Pasha, the Head of the Navy Manufacturing Commission Hüsnü Pasha, the Abdülhamid's First Secretary Tahsin Pasha, his Second Secretary Izzat Pasha, the Minister of the Navy Hasan Hüsnü Pasha were all among the members of this commission. At first the Minister of the Navy, Hasan Fehmi Pasha, presided over the commission. Later, when the Grand Vizier Mehmed Ferid Pasha got involved, the presidency naturally transferred to him.[4] As a matter of fact, Izzat Pasha had always been the commander in chief, right from the inception of the Hejaz railway idea. The *Komisyon-ı Ali* would meet in Izzat Pasha's suite in the palace, which indicated that he was the key figure of the Hejaz railway project.[5] According to Austrian ambassador Pallavicini, "the Damascus-born Izzat Pasha was the heart and soul of the commission and the railroad. With his loyalty to Abdülhamid and his great organizational skills, he had put his heart into the mission."[6] German consul Bodman was also to state that Izzat Pasha, thanks to his energy and political cunning, had eliminated the seemingly insurmountable problems one by one.[7]

The powers and responsibilities of the Komisyon-ı Ali in İstanbul and the local commission in Damascus were clearly defined. The general strategy would be determined by the central commission in İstanbul and the local initiative was delegated to the Damascus Commission. For instance, the powers and responsibilities of the Komisyon-ı Ali included the determination and execution of the funding strategy and mobilization of income resources. As such, developing a recruitment policy for the railway personnel, appointments to critical positions, importing necessary products and equipment, organizing tenders for the infrastructure investments and making agreements with contractors, keeping in contact with the companies that had previously been given railway concessions in the region, and similar activities can all be listed within the responsibilities of the central office.[8] Any other expense was subject to the permission of Komisyon-ı Ali.[9]

The Damascus Commission, on the other hand, was trying to realize the railway construction in accordance with the plans and projects coming from İstanbul. Using the money sent from İstanbul as determined by the Komisyon-ı Ali—in other words, paying the salaries and the money pledged to the contractorsand procuring everything required for the construction—these were all among the duties of the Damascus Commission.[10] The powers of the local commission were not broad enough to give it manoeverability, which prevented it from addressing urgent problems. In 1902, when it was realized that the local commission had almost no authority without the approval of the Komisyon-ı Ali, its powers were expanded. Henceforth the commission in İstanbul would be consulted only when financial problems and vital issues emerged.[11] The Beirut commission that started to function in 1902 under the provincial governor, was responsible for the construction activities along Haifa branch line.[12]

B) NEGOTIATIONS WITH THE DHP COMPANY

Before the commencement of the construction, Komisyon-ı Ali had decided to negotiate with the French DHP Company, which had operated the Beirut–Damascus–Muzeirib line since 1895, and if possible to buy the Damascus–Muzeirib section. Taking over the Damascus–Muzeirib line from the French would save 120 kilometers of Hejaz railway construction. Likewise, instead of starting from Damascus, the construction could continue from Muzeirib, a point on the Damascus–Muzeirib rail route, which was already in use.

However, during the negotiations the French company authorities were reluctant to compromise and demanded the exorbitant sum of

10 million francs. Thus, instead of buying the railway, it was decided to install a second line parallel to the Damascus–Muzeirib line. Meissner, who was soon to be appointed to the Hejaz railway as chief engineer, had also estimated the value of the line at 5 million francs and imposed on the Ottoman government the idea that constructing a new line would be more rational.[13]

The Ottoman Empire's decision to construct a line parallel to the Damascus–Muzeirib railway had elicited intense reaction from the French. In İstanbul, Deputy Ambassador Bapst officially protested against the situation. Bapst wrote in his protest letter—sent on November 20, 1901, to Tevfik Pasha—that the line which the Ottoman government was planning to construct would drag the Damascus–Muzeirib line to bankruptcy. According to Bapst this part of the French railway could by no means compete against the Hejaz railway, because in a railway project constructed partly by donations, the state had the leeway to reduce the transportation fees to a great extent. But this was not the case for the French company. The interest payments on the invested capital and other costs would render it impossible. Bapst defined the situation as unfair competition.[14]

The Ottoman administration ignored Bapst's protest and launched the railway construction from Damascus. Thereupon, DHP continued its protests via the French embassy and strove to shift the starting point of the railway from Damascus to Muzeirib. They found their justification in the concession agreement, according to which the Ottoman Empire had pledged not to give permission to any other company to construct a railway within this region. According to the French, the Sublime Porte was violating these terms and bringing their railway operation to ruin. The DHP asserted that it had not obtained any kilometer guarantees from the Ottoman Empire, so it was impossible for the company to compete against the Hejaz railway.[15]

The Ottoman governors, on the other hand, defended themselves by stating that they had not given any concession to a foreign company as stated in the agreement, thus there was no violation of the terms. The Ottoman side argued that among the Beirut–Damascus–Muzeirib railway agreement terms, there was no provision that said the state could not construct a railway in the region.[16]

The efforts of the French embassy did not yield any results. In 1903, the Ottomans had constructed a railway between Damascus and Dera, parallel to the Damascus–Muzeirib line. When the DHP could not prevent the Hejaz railway from commencing in Damascus, it began to demand compensation for the losses incurred.[17] French ambassador Costanz provided full support to the DHP. At the end of the negotiations—which lasted for years—the DHP asked the Ottoman Empire to either buy the Damascus–Muzeirib line for 7.5 million francs[18] or shore up their operations with a specific kilometer guarantee.[19] When the Ottoman Empire attempted to include the Rayak–Hama railway route into the bargaining process, the French clearly stated that "they would never agree on selling it". This was because they were already getting kilometer guarantees from this part of the railway.[20] However; the Ottoman administration had the legal right to buy the line by fulfilling certain conditions stated in the agreement.[21] The French not only ignored this situation, but also made new requests for compensation. For example, they demanded that the Rayak–Hama line be extended to Aleppo with a kilometer guarantee.[22] At this stage, the Ottoman governors agreed to pay 7 million francs only to the Damascus–Muzeirib part of the railway.[23] When the Grand Vizier Ferid Pasha notified the Ambassador that they were ready to pay an amount very close to the French offer, Constanz began to drag his feet and said that he—on behalf of his government—was not interested in the sale of the Damascus–Muzeirib line, but was interested in being granted the Hama–Aleppo concession[24]. Marschall, the German ambassador, stated in his report to his own Ministry of Foreign Affairs that the 7 million francs the Ottoman Empire was ready to pay was too much.[25]

Revisiting the French Foreign Minister's statements, the reason why the great powers supported their own citizens within the Ottoman Empire becomes clear. We know that the ulterior motive behind the railway investments was to create areas of influence within the empire. Hence French Foreign Minister Delcasse's statement that renouncing the concession would be detrimental to French interests and reputation in Syria.[26] Similarly Costanz, who had initially welcomed Ottoman offers, started to raise difficulties upon the warning of the French Foreign Minister,[27] which also corroborates this argument. Apparently, the Foreign Minister attached far more importance to France's influence in the region than the company's income statement. Cemal Pasha wrote in his memoirs that the French Foreign Minister had objected to the restitution of the Damascus–Muzeirib concession on the grounds that it would harm France's interests in Syria.[28] It is also very interesting to note that it was the French ambassador who conducted negotiations for a private French company on behalf of "his government" and that the French administrators considered this to be perfectly normal.

The negotiations between the Ottoman administrators and DHP authorities lasted for years, finally coming to an end in April 1905. According to the eventual settlement, the Ottoman Empire agreed to pay compensation of 150,000 liras—approximately 3.4 million francs—to the DHP Company.[29] The Ottomans also pledged to grant the concession of a prospective railway between Damascus and Aleppo.[30]

This conflict between the Sublime Porte and the French had stemmed from a legal loophole. When the Ottoman government granted the Beirut–Damascus–Muzeirib concession to the DHP Company it had in fact agreed, just as the French claimed, not to let any other company construct a railway in the region. However, when the agreement was signed, the legal problems that could arise in case the Ottoman Empire itself attempted to construct a railway were not considered. At that time the Ottomans did not have a history of constructing and operating state railways. And the only initiative, that of Abdülaziz, had been a failure. In this regard, the Hejaz railway was a new phase in Ottoman history and it brought about new and unexpected problems.

The amount of compensation that the DHP Company received cannot be justified by looking at the terms of the agreement. Because the Ottoman government had not given another company the authority to construct a railway, it was the government itself that decided to build the railway in question. However if the relevant contractual clause had aimed at protecting the DHP from competition, the Ottoman government's decision was contradictory to the spirit of the agreement if not its wording. Still, the Ottoman government could have circumvented the problem by paying a lower compensation. As a matter of fact, the Ottoman administrators were aware and kept emphasizing that they were in the right. For instance, Grand Vizier Ferid and Foreign Minister İsmail Hakkı argued, referring to the DHP, that "the aforementioned Ottoman company had no right to claim any damage and this was clearly safeguarded by the agreement".[31] Similarly, another report reiterated that the agreement had no provision that restrained the Ottomans from constructing a new railway in the region.[32]

Although, legally speaking, the Sublime Porte had the upper hand, they decided to pay the compensation nonetheless. The motivation behind this was more general than regional. Above all, the Ottoman Treasury depended on French loans. French administrators who were aware of this fact had asked the Ottomans to offer concessions wherever possible, primarily in railway projects. DHP's demands were listed as preconditions of a possible loan agreement. The Hama–Aleppo concession, for instance, was demanded with a kilometer guarantee of 13,677 francs. And if the annual revenue of the Damascus–Muzeirib line were not to reach 7,500 liras, then the Ottomans were requested to compensate for their loss in the form of guarantees.[33] As an example of demands that had no bearing on the railway, we can give Costanz's demand that all materials imported from Germany be replaced by French counterparts.[34] Thus French diplomats raised all kinds of demands, relevant or not.[35]

As we have detailed in the previous chapter, Abdülhamid wanted to complete the Hejaz and Baghdad railways, which he considered to be a parts of whole, at all costs. He was ready to make compromises to bring about his big dream. At the time when the agreement was concluded with the DHP Company, one of the preconditions for the resumption of Baghdad railway construction was to tone down the French opposition. In fact, French opposition against the Anatolian and Baghdad railways had previously been neutralized by offering concessions. DHP had obtained the Beirut–Damascus–Muzeirib line concession as a result of the French reaction against the

Eskişehir–Konya concession granted to the Germans. Indeed it was for the same reason that the French entrepreneurs who operated the İzmir–Kasaba line had acquired the right to extend this line to Afyon.[36]

C) THE LAUNCH OF THE CONSTRUCTION AND RECRUITMENT OF THE TECHNICAL STAFF

In order to collect preliminary information on how to construct the railway, two Anatolia railway engineers were sent to the region. Under the chairmanship of Hajji Muhtar Beg, a group of Ottoman engineers started to work on the delineation of the route in accordance with their suggestions.[37] Commander Şevki Beg, who had completed his education in France, also took part in this commission.[38] The fact-finding activities could take place only along the traditional hajj route for security reasons. Therefore, the fact-finding group joined the hajj caravan heading for Mecca from Syria.[39] This caravan was composed of the Surra regiment coming from İstanbul and the hajjis waiting in Syria, and it was protected by an infantry regiment, a mountain battery and numerous cavalrymen.[40] This was because the Surra regiment carried precious presents sent by the Ottoman sultans to the sharif of Mecca and the Bedouin sheiks. Drawing the Hejaz railway route along the hajj route had important benefits. First and foremost, the scarce water resources, which would be greatly needed during the construction process, lay along this pack trail. As a matter of fact, the Damascus–Medina hajj route was chosen mainly because of the availability of water resources.

Hajji Muhtar Beg presented all his conclusions, the profiles and the maps that he drew, in the form a report to the Komisyon-ı Ali. In his report there were suggestions as to where the Hejaz railway should be routed and detailed information about the geography, water resources, reservoirs, fountains and wells.[41] A narrow gauge (1.050 mm) was preferred in order to lower the cost. This way the compliance problem with the French narrow-gauge railways would be resolved. This was crucial, for the imported railway materials would be transported via the French-operated Beirut–Damascus–Muzeirib line.

The Hejaz railway would start from Kadem, located 1 km south of central Damascus. Apri Efendi, one of the qualified engineers of the Damascus government, had reported that the route passing from Hauran and Karak districts would enhance trade and settlement opportunities in these districts. Besides, because of ample water, stone, lime and similar material resources, the line could be constructed at minimum cost.[42]

The Hejaz railway construction was launched on September 1, 1990, the anniversary of Abdülhamid's accession to the throne. Soon after the construction process had begun, it was seen that it would take longer than planned. Due to inadequate plans and projects, some of the lines had to be removed and reinstalled. Foreign diplomats who had previously worked in the region stated that the main reason for the problems was the inexperience of the Ottoman government. For instance, according to some European consuls, despite all his goodwill and honesty, Construction Minister Kazım Pasha was damaging the construction works by his lack of technical know-how. The German consul in Damascus wrote that the technical group would have had a more favorable working environment if the powers of Kazım Pasha had been restricted.[43] Maunsell, the British military attaché, also agreed with the view that Kazım Pasha would not be able to function as the general director of the Hejaz railway construction. Also, in an article published in *Frankfurter Zeitung*, it was written that in the meetings held in the Yildiz Palace no one had an idea on how to progress. Therefore, the need for foreign technical personnel was further emphasized. In the meeting, the first agenda item was the need to appoint a foreign advisor from the other railway companies within the empire.[44]

We learn from another German newspaper that Otto von Kapp, one of Abdülhamid's advisors appointed to work on the Baghdad railway, emphasized in his report to the sultan that giving authority and responsibility to the foreign engineers was a precondition for success in the construction process. According to Kapp, due to incompetence and mistakes, some of the initial lines installed by native engineers were of no use.[45] It is evident that Ottoman engineers had no experience in railway construction.

The *Hendese-i Mülkiye*, the first educational institution to train engineers in the empire was only 16 years old.[46] On average, 11 students had graduated from this school each year between 1884 and 1909. Some of the first Ottoman engineers with a four-year degree had served on the Baghdad and Hejaz lines and as officials of the Ministry of Public Works in the irrigation of the Konya plain.[47]

Once it became apparent that the objective of recruiting only Ottoman engineers was unrealistic, it was decided to appoint an experienced chief engineer, who had proven himself in his professional career, to lead the technical staff. In September 1900, the German engineer Heinrich August Meissner was appointed. Consequently, the number of foreign engineers —the majority of them German— started to increase.[48] Some of these foreign engineers had worked before on railways constructed by Germans in the Ottoman Empire.

The Ottoman government preferred German technical staff after having admitted that recruiting European personnel was inevitable. Continuous improvements in Ottoman-German relations could account for this. Abdülhamid still had no confidence in other European countries, especially Britain.[49] As already discussed in the first chapter, Germany was the only country the Ottoman Empire could rely on in order to survive by capitalizing on the conflicts amongst imperialist states. In 1904, twelve German engineers were working on the construction as opposed to one Greek, one Belgian, two Austrian, five French and five Italian engineers.[50] In 1906 Marschall wrote to Bülow that Izzat Pasha had told him that they were looking for one senior engineer and fifteen department chiefs. His report also emphasized Abdülhamid's preference for German engineers. Thus Tevfik Pasha, who was in Berlin, was authorized to run the process.[51]

Meissner had graduated from the Technical Faculty in Dresden in 1885 with exceptionally high grades. When he was in Dresden he had learned Turkish —a move which showed that even then he must have been planning to work on the Ottoman railways.[52] Thus he fulfilled his vision when his uncle Victor Tridon who worked as an engineer in İstanbul invited him. In his first experience in the

Ottoman Empire, Meissner started to work as an engineer on the İstanbul and Rumelia railways.[53]

Meissner, after having worked on railway technologies at a technical school in Prague, returned to İstanbul in 1887. On his second arrival, he worked as the chief engineer on the İzmit-Ankara and Salonika-Manastır railways, from 1889 to 1892 and 1892 to 1894, respectively. Between 1894 and 1896 he worked as the chief engineer on the seventh section of the Salonica-Alexandroupolis line. In 1896 Meissner started to work as consultant for the Ottoman railways in İstanbul. In September 1900, in line with the developments mentioned above, the technical management of the Hejaz railway was handed over to Meissner as the chief engineer.[54]

Meissner obviously boasted all the qualities that the Ottoman government had stipulated for this post. In addition to being a German citizen, he also had vast experience on the Ottoman railways. He was familiar with the circumstances, traditions and people of the country and he spoke Turkish. These were all great advantages. Meanwhile, Abdülhamid must have been very pleased by Meissner's achievements, as he was granted with the title of Pasha on March 7, 1904.[55] According to Auler, another German pasha who worked in the Ottoman army, Meissner Pasha was someone who had enhanced his deep knowledge through his experience, had spent most of his life in the Ottoman Empire, could speak Turkish fluently and saw eye to eye with the railway personnel and the public.[56]

Foreign supervisors wrote that, with his competence in English and Italian in addition to Ottoman Turkish, Meissner could manipulate Ottoman administrators who lacked technical know-how. For instance, the English military attaché Maunsell highlighted Meissner's mastery in undermining the authority of Ottoman government officials over him. Although he had initially gotten into conflicts with the Hejaz Railway Construction Director Kazım Pasha, he later managed to prevent Kazım Pasha and anyone else from the Damascus Commission from interfering with his work.[57] His salary had been raised to 1,000 liras. Besides, Abdülhamid had given him vast lands in Maan located on the 460th kilometer of the Hejaz railway and permitted him to build a mansion there.[58]

Abdülhamid insisted on employing Ottoman engineers on the Hejaz railway although they were inexperienced. In 1903 seven Ottoman engineers were working in Mecca; in 1904 this number increased to 17.[59] As the objective was to increase the number of locals, every year a certain number of Ottoman engineers were sent to Europe in order to develop their professional skills. These engineers would not only learn western technology but also acquire western languages. Nazif Beg el Khalidi, who was one of these engineers, came from one of the prominent Sunni families in Jerusalem. After graduating from Hendesehane-i Mulkiye in İstanbul, he went to Paris and completed his education at the Ecole Polytechnique. On his return, he started to work on the Hejaz railway and led the construction of the Damascus train station and several tunnels and bridges. He also participated in the railway construction work in Amman.[60] The Islamic nature of the project was not the only reason behind the efforts to educate the Ottoman engineers. Non-muslims were forbidden to enter the lands between al-Ula and Medina, in the south of the region, through which the rails would pass. Thus, this section had to be built by Muslim engineers.[61] As a matter of fact, in one of the declarations of the Ministry of War, the importance of this issue was emphasized:

> ... Although there should have been no Christians working on the railway line from Al-Ula to Medina, there were Christians bearing Muslim names. The Arab sheikhs stated that this would cause problems, and in order to avoid these problems, such people had to be sent back where they came from.[62]

Haji Muhtar Beg was appointed as Meissner's assistant and his exclusively Muslim staff successfully accomplished this task.[63] Haji Muhtar Beg successfully led the construction of the bridges in the Yarmuk valley, as well as drawing the maps of the railway route and the Amman substation.

Auler Pasha, who made an inspection trip to the Hejaz railway in 1906, reported that the number of Ottoman engineers had reached 25. According to his remarks, within the same period ten foreign engineers were working on the railway.[64] Apparently, more and more Ottoman engineers were being recruited. The fact that this issue was frequently discussed by the Komisyon-ı Ali demonstrates the importance the Ottoman government attached to it.[65] For example, in 1903 half of the Hendesehane-i Mülkiye graduates were to be employed on the railway. An engineer who refused to take on this duty would never be able to work in public service again. Galip Effendi, who was one of the railway engineers and who had come to İstanbul without permission, had to "be sent back to his official post immediately". Accordingly, the engineer who was sent back to Damascus would be considered to have relinquished his post on the Hejaz line. For fear that Galip Efendi would set a negative example for others, he was dismissed from public service altogether.[66]

The Ottoman engineers' performance was as anticipated. In fact, Kapp von Gültstein, who was one of Wilhelm II's counselors and was appointed to the Hejaz line as an advisor, praised the Hendese-i Mülkiye graduates for their work. According to him, the Hejaz railway under the control of Meissner was a marvelous opportunity for training Muslim engineers.[67] Thus Mustafa Şevki Atayman, who graduated from Hendese-i Mulkiye in 1897, would later on, between 1914–18, work as the operations department manager on the Hejaz railway.[68]

Because of the exceptional performance of the Ottoman engineers, the obligatory service responsibility reserved for Hendese-i Mülkiye graduates was extended to the graduates of *Mekteb-i Sanayii* and *Darü'l-Hayr-i Ali* in 1907. "Upon the appeal of the Komisyon-ı Ali, an Imperial Decree was issued on 20 February, 1907, which stated that the students who graduated from Darülhayr-i Ali and Mekteb-i Sanayii with high degrees would be employed on the Hejaz railway with their term of office being exempt from military service on condition that they served five years."[69] And in January 1909, five graduates from Darülhayr-ı Ali were employed on the railway.[70] In addition, railway courses were added to school curricula.[71]

D) PROVISION OF RAILWAY MATERIALS

The initial idea was to produce the railway materials in Ottoman workshops, due to the Islamic nature of the Hejaz railway. The 1,382,000 traverses that would be needed in the first stage would be cut from the stateowned forests and the Komisyonu-ı Ali decided to collect used iron from all provinces to use in rail manufacturing. On the other hand, as the commission was aware that the iron produced in the *Tophane* (foundry) and *Tersane* (shipyard) would not be sufficient, it proposed to import 10,000 iron bars by tender.[72] Although the production started at Tersane-i Amire (Naval Arsenal) in the Golden Horn, the rails manufactured were not hard enough to carry the trains and, consequently, rail were imported from Belgium and Britain. America was another country from which rails were imported. Belgian rails were used between Damascus and Maan and American ones were used from Maan towards the south.[73]

Similarly, because the traverses manufactured from the wood of Anatolian trees could not resist the hot climate of Arabia, traverse importation became necessary. In order not to import steel traverses, the Hejaz Railway Administration had tried for three years to produce wooden traverses that would function the same.[74] The plan to supply traverses by tender from domestic producers failed. For example Mehmet Beg from the Milas gentry, who had won the tender, could not fulfill the commitment and another man named Mustafa had been put on trial on charges of bribery.[75] In the end, when wooden traverses installed up to the Zarqa station performed poorly, it was decided to use only imported steel traverses. Furthermore, the wooden traverses that were installed up to Zarqa had to be removed and replaced by steel rails. Steel was imported from a factory in the Oberhausen region in Germany.[76] Another factory from which steel rails and traverses were supplied was the Düsseldorf Stahlwerks–Verband. An agreement for 22,000 tons of traverses was signed with this company and trade continued in the coming years.[77]

In the end all the railway materials had to be imported. Political and economic relations between the Ottoman Empire and Germany and the chief engineer Meissner's influence led the majority of the imports to be made from Germany. For instance, pumps and boilers came from the Weise und Monsky, Korting and Reinsch factories located in Halle, Hannover and Dresden; assorted military equipment came from Berlin Koppel; and tanks and coaches came from Frankfurt Süddeutsche Waggonfabrik.[78] All but six of the locomotives had been manufactured in Germany.[79] Munich Krauss, Düsseldorf Hohenzollern, Chemnitz Richard Hartmann, Kassel Henschel und Sohn, Kirchen Jung, Halle Lindner and Gotha Rothmann were among the factories that exported locomotives for the Hejaz railway. Only one passenger wagon and a wagon that was designed exclusively for prayers were built in the *Tersane-i Amire*.[80]

When diplomatic documents about the Hejaz railway are analyzed, it becomes clear that German capital and the German government were undoubtedly intertwined. For instance, in 1910 Stahlwerk could not deliver the rails and traverses that the Hejaz Railway Administration had ordered. Thus, in order to avoid paying compensation, the company asked the German embassy to put pressure on the Hejaz Railway Administration.[81] When another company, Deutsche Levante Linie, heard that the Afula–Jerusalem line would be constructed in 1912, it requested the Ministry of Foreign Affairs to examine potential opportunities that this new construction could offer to the German industry.[82] The reason behind this shipping company's involvement was the fact that it had transported the exported materials to the region.

The rails and traverses imported from Germany to the Ottoman Empire between 1910 and 1913 were worth more than 19 million marcs and, as is seen in the above example, the core activity of some German shipping companies was transporting railway materials.[83] The German Foreign Ministry asked the Haifa Consul Hardegg to write a report on the possible advantages the railway construction would offer to German industry. Hardegg prepared a report which stated that as the Ottoman government would import construction materials, German industry had to seize the opportunity. The consul also remarked that the Komisyon-ı Ali would be the ultimate decision-making authority on the quantities, so he could not do anything else in Haifa: the responsibility lay on the shoulders of the diplomats in İstanbul to influence this process. Besides, the German operating manager of the Hejaz railway, Dieckmann, had asked him to learn the prices of railway material manufactured in Prussian State Railroad workshops. That's why Hardegg wanted the prices to be reported to him.[84]

Two days later Hardegg suggested in his report to give state subsidy to two German companies named A. Dück & Co. and Schmidt & Co. that operated in Haifa.[85] In his report to the chancellor only three days later, he presented a detailed examination of the amount of German-made railway materials that the Ottoman Empire had imported for the Hejaz railway.[86] In another report that he prepared for the Haifa–Acre extension, he stated that two iron bridges were constructed by German companies.[87]

The importation of materials for the Hejaz railway was a strictly regulated process. Firstly, a call for tenders was published in the newspapers. From time to time various Ottoman authorities would make sure these tenders were announced to the widest audience possible. In one of its declarations the Directorate of the Press Department (*Matbuat-ı Daire*) noted that as the materials in question would mostly be imported from Europe, tender calls had to be issued in several languages and newspapers.[88] Eventually, an agreement was signed with the company that made the optimal offer. Later, wagons and similar import materials would be checked by three Ottoman engineers at the production phase.[89]

E) HAIFA–DERA BRANCH LINE

1- *The Construction of the Line*
In the beginning the materials imported for the railway were being transported by the Beirut–Damascus–Muzeirib line operated by DHP. This resulted in soaring transportation costs and gave the French company political leeway.[90] There had already been a serious controversy between DHP and the Ottoman government. In order to elliminate such problems, the Ottoman government decided to instal a branch line from Dera, located on the Hejaz railway, to Haifa on the

Mediterranean coast. This would enable them to transport the imported railway materials free of charge and block French interference. The branch line would also provide an exit from the Hejaz railway to the Mediterranean. However, as we know, the railway concession in this region had been given to a British company, the Syria Ottoman Railway Company. The concession of this company, which failed to fulfill its obligations, had been nullified by the Ottoman authorities.[91] In fact, the company could not overcome the challenges and the construction could only continue for 8 km. By paying 155,000 pounds to the company, which was in financial difficulty, the Ottoman government both annulled the Haifa–Damascus concession and acquired the 8 km line.[92]

The Haifa–Dera line construction began in April 1903 and was run by German engineers. For instance, the operation in Jordan was led by Chief Engineer Weiler.[93] The wagons and locomotives were imported from Germany; the rails and traverses were imported from America and Belgium.[94] Two used locomotives were bought from DHP and Abdülhamid had a saloon coach built in İstanbul which he donated to the railway administration.[95] Apart from the Germans, several Austrian and Italian engineers and a few Ottoman engineers were employed.

Contractors were commissioned at infrastructure activities such as bridges, channels, tunnels, water supplies and station buildings. Infrastructure between Haifa and Jordan had been contracted to a group of engineers composed of two Germans, one Italian, one Austrian and one Ottoman from Damascus. It is presumed that the contractors used civil servants and wage labor and made a profit of 10–12 per cent in the region.[96]

There were important technical problems associated with the Haifa line. For instance, in the Yarmouk plain section, eight tunnels, 83 bridges, 246 aqueducts and many headwalls and viaducts had to be constructed. Later on, construction had to continue at an altitude of approximately 600 meters in the Hauran plain. Although stone was used in the construction of bridges in the main section of the Hejaz railway, iron bridge construction was required in four places.[97] Despite all these challenges, in late 1904 the main line had reached Muzeirib.

Although infrastructure work had not been completed as planned, the Komisyon-ı Ali decided to open the Haifa–Dera line to passenger and freight traffic as soon as the rails were installed. Since June 1903, 76,618 tons of construction materials had already been transported via this line.[98] The idea was to immediately start benefiting from the operating income and carry on the construction. But eventually the rails were damaged; for example in 1906, due to flooding in the Yarmouk plain, the line had to be closed temporarily.[99] Every day a train from Haifa to Damascus and one from Damascus to Haifa departed. The journey between Haifa and Damascus lasted approximately 11.5 hours.[100]

The report of the Komisyon-ı Ali dated July 18, 1904, emphasized that the Haifa line was the most challenging one. It was also stated that even big businesses constructed only about 120 km annually, and so the construction of 550 km within three years was to be celebrated as a great success.[101] As a matter of fact Schroeder, the German consul in Beirut, confirmed the opinion of the Commission members and hailed the Haifa–Damascus line as a tremendous technical achievement.[102] Despite the resistance of the Bedouin sheiks, the Haifa line had been completed and the traditional camel transportation in the region had been negatively affected.

2- *The Economic Dimension of the Haifa Railway*

Among all the parts of the Hejaz railway, only the Haifa section was constructed with economic aspirations. Indeed, regions such as Hauran, which constituted the hinterland of Haifa, and the Jordan Plain were the most fertile agricultural lands of Arabia. Before the construction of the railway, the crops grown in the region were transported to the coast, Haifa or Acre by camel. It had been said for many years that had there been faster and cheaper transportation between the hinterland and Haifa, agricultural production and Syrian exports would have boomed. In fact, completion of the Haifa line brought dynamism to the region's economic life.[103] In 1913 the German consul in Haifa wrote that the Haifa line accounted for more than one-third of the passenger and freight traffic of the Hejaz railway.[104]

The connection of Syria to its hinterland by railway brought about rapid economic and social development in Haifa; however the development of the neighboring city of Acre was inhibited. For instance in 1911, while Haifa's population was 23,000, the population of Acre dropped to 12,000. However, in 1868 while 15,000 people had lived in Acre, the number of inhabitants in Haifa had been only 4,000.[105] From 1904 to 1913 the volume of shipping operations from Haifa had increased from 310,103 tons to 810,418 tons. In the meantime shipping operations at the Acre port had shrunk from 96,458 tons to 26,418 tons, a two-thirds decline in total.[106] Many shipping agencies were established in Haifa and the banking sector flourished. Apart from the Ottoman Bank, the Deutsche Palaestine Bank and the Anglo Palaestine Bank were operating in the city.[107] By the way, we should state that the number of people who arrived at Haifa by sea and continued their journey by the recently-inaugurated railway was increasing day by day. In the beginning, Christian pilgrims and tourists were mostly using the Haifa railway to visit Jerusalem and its vicinity. With the efforts of the Ottoman government, Muslim hajjis trying to reach Medina and Mecca started to use the Haifa line too. Thousands of Muslims, especially from Egypt, would come to Haifa by sea and from there set out for the holy land by train.[108]

Soaring passenger and freight traffic and increasing export and import activities via the Haifa line made it obligatory to construct a port in the city. Since there was no port in Haifa, passengers and freight had to be transported to the coast by boat. This became a challenge at times when the boats could not come aboard due to stormy weather.[109]

Berliner Tageblatt suggested two reasons no port had been built in Haifa thus far. First, Ottoman rulers refused to grant the Haifa port concession to a foreign company. The foreign involvement at a very critical point on the Hejaz railway was not welcomed by the Ottoman administrators, as the independence of the railway was crucial. This could bring harm to the Islamic character of the line.[110] Also according to the Austrian ambassador, the Ottoman government considered the operation of the Haifa port by foreigners detrimental to their reputation, for this port would function as the terminus of a railway used by hajjis.[111] In fact the Ottoman administration had decided to construct a state-owned port in Haifa. The government took measures to prevent the landlords from expanding their assets towards the coast as they would probably aspire to partake in the unearned income from port construction. The cautiousness of the government shows how committed they were.[112]

According to *Berliner Tageblatt*, another obstacle before the construction of a port in Haifa was the reaction of the populace in Acre. According to these people, the construction of a port in Haifa would mean the downfall of Acre both socially and economically.[113] This was emphatically emphasized by the notables of Acre, echoing the views of the *Berliner Tageblatt* author, in a telegraph that they sent to the central office in May 1912.

> Rumour has it that the port to be constructed by the Hejaz railway will be constructed in Haifa upon German request. Haifa is a city surrounded by mountains and sandy places. It lacks the surrounding to which it can shed the light of two important sources of happiness; the railway and the port. It is subject to strong northern winds. Acre, on the other hand, is a city neighboring a fertile plain and surrounded solely by Ottoman settlements. It is a city that has a significant historical background, thus with the completion of a 14 km line it may serve as the beginning of the Hejaz railway. [...] We would like to be sure that, given that it is possible to construct a magnificent port in Acre at half or one-quarter of the cost that will be spent in Haifa, and given that this construction will have greater benefit for the people both financially and politically, the Ottoman Grand Vizierate will not let the Ottoman Empire and the Ottoman treasury be squandered due to personal interests of the Germans and the Jews.[114]

A report written by the German consulate in Haifa in 1914 gives information about both the socio-economic development of Haifa and Germany's close interest in this region, somewhat proving the inhabitants of Acre right. The title of the report sent to Chancellor Hollweg was "German interests related to the Hejaz railway" and the plenitude of notes taken by Emperor Wilhelm while examining the reports indicated that greater emphasis was put on reinforcing the German influence in the region.[115] Hardegg expounds on the rapid development in the city that started with the launch of the Haifa line. He writes that, after the construction of the railway, the population of Haifa tripled and customs duties increased five-fold, the economic life—which was generally in the hands of the Germans—was revived, and the price of land possessed by the German colonizers doubled: "15 years ago one hectare of land was approximately 100–200 francs, today the price is approximately 350–500 francs. Meanwhile land prices have risen by eight to ten times."[116]

According to Hardegg the changes that the Hejaz railway brought were in the interest of the German colony in Haifa. Moreover, since the beginning, German citizens had always earned higher incomes by assuming higher positions throughout the construction process. The consul wrote,

> Including the construction manager and the directorate of the technical board, eight German subjects currently work on the Hejaz railway. Owing to the Germans serving as senior officials, the name "German" has gained prestige and German industry has received large orders. Consequently German industry has exported 20 million francs worth of railway materials to the railway. The exports to the region along with the railway have also invigorated ship traffic in Haifa.

It is interesting that the consul in Haifa continued his report with calculations concerning potential hajj traffic once the Hejaz and Baghdad railways were completed. The consul's elaboration of such details was due to the German interest in the region and the idea that the best way to compete against other imperialist states was to see eye to eye with the Ottoman Empire. According to Hardegg's calculations, approximately 20,000 hajjis would use the railway and, assuming the tickets were 3 liras, expected annual income was 60,000 liras or 1,104,000 marcs. This amount could be increased further by introducing several arrrangements. The Germans had to keep giving guidance to Ottoman authorities to ensure maximum yield. At the end of his report, the consul stated that the Hejaz railway region would become a market for an array of German products. Postcards showing the sacred places, guide books about the holy places, water bottles, colored clothes, paper, glassware and canned goods were examples of the products for which potential demand could be generated.[117] Apart from the consul in Haifa, Auler Pasha also thought that the strongest and most influential group among the Europeans in the city was the Germans.[118] The formation of the aforementioned German colony in 1868 had started by the migration of Pietists, a 17th century religious movement originating in Germany in reaction to formalism and Jews from Württemberg to the region.

Given the fact that German influence on the Ottoman Empire had grown tremendously, German administrators' interest in Haifa becomes fairly understandable. Through the Anatolian Baghdad railways, German influence was expanding from Anatolia to Arabia. The Hejaz railway had created a new opportunity for the Germans to penetrate even deeper. The Baghdad and Hejaz railways were parts of Abdülhamid's railway project and, when the decision not to employ foreign personnel on the Hejaz line was suspended, it was anticipated that Germans would come to the forefront in the construction of the Hejaz railway. Capitalizing on the ongoing Ottoman–German rapproachement, Germans would surely try to seize this opportunity in Syria, where they had always remained in the background compared to the French and British, and they would be on the prowl to increase their influence in the region. That is why Germany had endorsed the Hejaz railway idea right from the beginning.[119] The Germans welcomed the Hejaz railway with regards to its military benefits as well. If the Baghdad and Hejaz railways could be joined at Aleppo, then it would be possible to attack the British both from the Red Sea coasts and the Persian Gulf.

3- *Rivalry against DHP Company and the Pro-Hejaz Railway Attitude of the Germans*

It was clear that the Haifa–Dera–Damascus line would compete with the DHP Company, which had the same passenger and freight potential. These two railways connected the Mediterranean coasts and the fertile lands of Syria and ran parallel to each other from Damascus to Muzeirib with a distance ranging from 2 to 15 km in between. Although the Haifa–Damascus line was more than 137 km, it had an important advantage. The Haifa line was partly financed by the state and through donations. This gave the railway administration flexibility in setting passenger and freight tariffs—which was not the case for the French company, which had to pay the interest on the invested capital.[120] Due to the effects of the 1908 strike wave, the French company had to raise the salaries of the employees by 50 per cent; another economic challenge they had to endure. Besides, the Beirut Port fees were kept high due to its monopoly and this fact had started to affect the DHP negatively. Thus the Hejaz Railway Administration had eliminated the French line's monopoly on transportation and could take the passenger and freight traffic, though partly, to Haifa.[121]

However this issue was blown out of proportion by the press. In order to understand the extent of the rivalry, it should suffice to examine the transportation statistics. In 1911, 48,863 tons (91 per cent of the total amount) of freight was transported from Beirut to Damascus. On the other hand, 4,935 tons (9 per cent of the total) of freight was transported from Haifa. In the following year we see the situation change in favor of Haifa: 8,952 tons (21 per cent) from Haifa to Damascus and 34,095 tons (79 per cent) from Beirut to Damascus. We also see Beirut maintaining its importance in freight shipment from Damascus to the Mediterranean coast. Whereas in 1911 36,300 tons of freight, which equals 96 per cent of the total amount, could be transported from Damascus to Beirut, Haifa's share was only 1,567 tons, 4 per cent of the total freight. In 1912, these become 38,813 tons (92 per cent) and 3,560 tons (8 per cent) respectively.[122]

As is seen, despite discounts on ticket fees and low import and export costs, the Hejaz railway fell short of expectations in terms of shifting traffic to Haifa. The DHP could survive thanks to the institutionalization of French economic dominance, which had historical roots. The banking and trading capital that connected Damascus and Beirut was under French control. The French, who were in the saddle in the Damascus–Mediterranean trade, were more scared of losing their superiority between Damascus and Muzeirib. For this reason, in order to prevent the shift of products of the new Hauran region to the Haifa line, they offered discounts of up to 50 per cent on the transportation fees between Damascus and Muzeirib.[123] Thus price wars started between the Hejaz railway and DHP.[124]

In 1912, reducing the freight and passenger fees to below DHP levels gave the Hejaz railway significant leverage in the export of agricultural products from the Hauran region. For instance, DHP's share in freight transportation between Damascus and Muzeirib was 66,205 tons, hence 65 per cent in 1910, that of the Hejaz railway was 36,303 tons and 35 per cent. These figures changed rapidly and remarkably in favor of the Hejaz railway in 1911 and 1912. In 1911 the two companies were neck-and-neck in transportation activity in the region, with approximately 40 tons each. However in 1912, as a result of huge discounts the Hejaz railway offered on its fees, its share increased to 65 per cent—96,671 tons—whereas DHP's share plummeted to 35 per cent, 52,839 tons.[125] In 1912 the gross income of DHP had decreased by 135,366 francs with respect to the previous year. But apart from the competition, the quarantine precautions taken due to a cholera epidemic also had a role in this decline. The same precautions had affected the Hejaz line revenues negatively as well.[126]

The following year, the Hejaz Railway Administration had to step back from its price-cutting policy.[127] However this decision was not related to the competition between DHP and the Hejaz Railway Administration. In fact, Hejaz railway administrators had to give in to French diplomatic pressure. It was after one year that France gained the opportunity to enhance its economic and political power in Syria through diplomatic mechanisms. In 1913, when the Ottoman Empire turned to France for a new loan agreement, the French understood that the Sublime Porte had its back against the wall and they knew exactly how to capitalize on it. They raised entirely irrelevant demands, such as the entitlement to control and act autonomously at their community schools within the Ottoman Empire.[128] As a precondition of a 700 million franc loan agreement, in addition to their demands to strengthen their influence, they also conveyed their requests concerning DHP.[129] On November 6, 1913, at a meeting in the presidential palace attended by French government members and diplomats, it was decided to enhance French railway concessions in Syria.[130]

Once again the Ottoman government had to accept all French demands as they could not extricate themselves from their financial predicament. Among these demands were the extension of existing lines like Jaffa–Jerusalem and new railway concessions such as Samsun–Sivas. The concessions to construct and operate the Haifa, Jaffa and Tripoli ports were also granted to the French.[131] Although DHP had received compensation in return for the losses it suffered due to the Hejaz railway in 1905, it also received the Ottoman Empire's guarantee of annual revenue injection. Moreover, the Hejaz Railway Administration promised not to reduce the transportation fees below a floor price.[132] The French demands in return for extending loans were not without precedent. In 1904, the French Foreign Minister had demanded authorization from the Ottoman government to construct the Hama–Aleppo railway. And not surprisingly, this time in 1913 when Cavit Beg was in Paris to sign a loan agreement, France raised difficulties because of the Syrian railways.[133]

By the way, we should also state that the inhabitants of the region had concerns about the concessions granted to the French. As mentioned before, granting Haifa port concessions to foreigners was still a sensitive issue. For instance, the Muslims of Haifa had sent a telegraph to the Grand Vizier protesting the extension of port concessions in their city to foreigners. The telegraph also emphasized their concern that the Hejaz railway would be at risk.[134] The Sublime Porte was not indifferent to the Muslim protest in Haifa, and sent a telegraph giving assurances to the residents of Haifa that the Hejaz railway would absolutely be constructed without foreign control.[135]

Furthermore, the Ottomans had fallen out of grace in the eyes of the Arabs in the aftermath of the Turco–Italian War.[136] Arab nationalism was on the rise in Syria. Parallel to increasing French influence in the region, the pro-British tendencies among the Arab nationalists had been replaced by pro-French sentiments. France, through its railway and port concessions, was trying to spread its influence from Syria towards Palestine, opening churches and monasteries and funding Maronite schools.[137] Miquel, the German consul in Cairo, made interesting remarks about the Arab nationalists' pro-French attitude. According to him, the Syrian nationalists who were seeking political autonomy were pleased with the revival of trade and economy in their region owing to the Hejaz railway. In other words, the change in their attitude arose not from fellow-feeling towards the French, but from financial interest. According to Miquel, another factor that drew the Syrian people away from the Ottomans was their belief that the Ottomans could no longer provide them with the security they needed.[138]

Miquel had talked to one of the prominent nationalists of the Syrian Congress. This secret member of the congress had told him that Syrians were not pleased with the status quo, and they wanted to change it. This change could be the establishment of autonomy like that of Lebanon, without seceding from the Ottoman Empire. If this could come true, then an agreement with a foreign power could be considered. Another Arab congress member made this rather interesting statement: "The heart and mind of Syria was yet to be won. As the Syrian relations with the Germans, Russians and Italians were limited, the only countries left were Britain and France. The French trump card was the Maronites, with whom they were still related, and likewise the trump card the British wielded was the Druze."

In his report Miquel was also referring to the "media war" that the British and French in Syria had launched in order to win the Syrians. Accordingly, *al-Muayyad*, the semi-official media arm of the British, was supported initially by Lord Cromer and then by Kitchener for propaganda purposes. The newspaper was incessantly reporting about the French administration's despotism in

Damascus, Algeria and Tunisia and offering "the tolerant and welfare-oriented British administrative mentality" in Muslim countries as an alternative. On the other hand, the French media outlet *Journal du Caire* published a series of articles titled "Lettres de Syrie" with propaganda purposes.[139]

The agreement signed with the DHP Company did not enter into force due to World War I. The French would have tried to acquire the Haifa line, were it not for the war.[140] For instance, it was rumored that the line would be rented by DHP for 75 years in return for 120,000 francs. Another alternative was to transfer the line to a French/Ottoman joint administration chaired by a French authority.[141] The French considered the Haifa–Dera railway in Syria, which they defined as their backyard, to be a deployment mechanism for the Ottoman/German powers. What is interesting is that the Germans, as if proving the French right, did actually see the railway as such. Hardegg, the consul in Haifa, being more royalist than the king, stated that the Germans had to fight back at every step the French made towards the Hejaz railway. His explanations demonstrated how complex the imperialist battle fought within the Ottoman Empire was. Hardegg continued, "Haifa, where German influence was strongly felt, had blossomed thanks to Germans. But in one single move the French wanted to halt the development of Haifa and elevate Beirut which was under their influence."[142]

Another report by Hardegg welcomed the collaboration between Haydar Beg from Syria and Deutsche Orientbank to gain new concessions as a positive move.[143] Ambassador Wangenheim was very vocal: "Although apparently they have nothing to do with us, we should keep an eye on French demands. The Hejaz railway is massively important for us and thus we support the railway in every way. But these demands are threatening the Hejaz railway."[144]

Seeing the Hejaz railway as a German response against French attacks was not peculiar to diplomats. The German media covered stories along these same lines. For example in a series of articles published in *Schlesische Zeitung*, the "frenchification of the Hejaz railway to which Germans put their labour" was deemed unacceptable and something to be resisted. French initiatives targeting Haifa—the port concession being first among them— were perceived as a conspiracy against Germany, as if these territories were not part of the Ottoman Empire. In an article titled "Französische Plaene in Syrien und Palaestina" it was stated that the French entrepreneurs had "considerably increased the French influence against Germans in Palestine". According to the author, just like the French, a German company should have applied for the Haifa port concession. This application could become one of the most effective instruments of struggle against French authority in Syria and Palestine. "Unfortunately" there was no such application. French shipping companies were regularly cruising the region twice or three times a week and returning to Marseille and Italy full of cargo. On the contrary the only German shipping company in the region, Deutsche Levante Linie, did not have regular operations. Life became harder after the French had gained the Haifa port concession. At the end of the article the author asked, "When shall we protect the German influence in Palestine from French attacks?" Part of the answer was taking steps to spread German propaganda in schools and encouraging use of the German language. The day before, the same newspaper had analyzed one by one all the railway and port initiatives—some of which could be realized, like the Jaffa–Jerusalem of the French in Syria and Palestine, and some which were planned but could not be realized, like the Jaffa–Port Said.[145]

The Ottoman Empire had sensed the possible pressure that would come from the French on selling the line to them, so in January 1914 the empire tied the whole Hejaz railway to the Evkaf Ministry (Ministry of Pious Foundations),[146] thereby according the railway foundation a status which prohibited its being sold.[147] In any case, after the outbreak of the war, the Hejaz line would be connected to the Ministry of War.[148]

F) INAUGURATION OF NEW LINES (1904–08)

Parallel to the inauguration of the Haifa–Dera line, the main line of the Hejaz railway was proceeding from Damascus to the south, towards the holy cities. Primarily the 13 km part between Muzeirib and Dera was constructed and it was partly opened to traffic on September 1, 1901, in order to contribute to the financing of further construction. The negotiations with DHP actually impeded the work, and the construction continued only towards the south. As we know an agreement reached with the French would save the railway administration from constructing the part from Damascus to Muzeirib. On September 1, 1902, the construction reached Zarqa, located 80 km from Dera.[149] When there was no hope left of making an agreement with the French, in 1903, the 123 km section between Damascus and Dera was completed. The same year the railway construction continued from Zarqa towards the south and the section between Zarqa and Quatrana was constructed (124 km). Both of them were launched on September 1, 1903. From then on, the opening ceremonies were to be performed on September 1, the anniversary of Abdülhamid's accession to the throne.

In 1904 there were two trains shuttling from opposite directions between Damascus and Amman. The train that departed at 8:00 a.m. would arrive in Amman at 9:00 p.m.[150] The same year, in 1904, the construction reached Maan, located on the 460th kilometer of the railway and five freight trains and two passenger trains started running every week.[151] On September 1, 1904, the first official inauguration ceremony was held. From that time on, the Ottoman administration would pay a great deal of attention to the inauguration ceremonies and use them for propaganda purposes. This way, the message that the Hejaz railway was not a fantasy and that they were reaching their objectives gradually could be conveyed. It provided an opportunity to show people that their donations were not in vain. When in 1904 engineers and employees could not be paid and the construction came to a halt, it was decided to open a credit account at Ziraat Bankası in order to "avoid unfortunate situations that could lead to negative rumors within the Islamic World".[152] Overcoming challenges not only allayed the doubts of Muslims but also impressed the orientalists, who had deemed the Hejaz railway unfeasible, to change their minds. For example, Eduard Mygind admitted that his negative assertion about the feasibility of the Hejaz railway was wrong.[153]

Abdülhamid attached great significance to the inauguration ceremony of the Maan line. He established a committee and appointed Turhan Pasha as its head. In telegrams sent to the districts, they asked local notables to send Izzat Pasha a list of those who would like to participate in the ceremony.[154] It was not only the Ottoman elite that was invited to the celebrations. High-ranking officials and journalists from various countries were also invited and all their transportation and accommodation expenses were covered by the Ottoman Empire. Eduard Mygind from the newspaper *Berliner Tageblatt* and Auler Pasha, who was working in the Ottoman Empire, were also on the guest list.[155] Auler was received with a very warm welcome as being "Allemania Pasha".[156]

Nearly 50 bureaucrats and journalists, among them Rahmi Pasha who was the first aide-de-camp of Abdülhamid, Cevad Pasha, Mustafa Effendi from the Ministry of Public Works, and legal adviser Mehmet Ali Beg, who was the son of Izzat Pasha, had started their journey from İstanbul by a vessel named *İzmir* and arrived in Beirut on August 23. Among the passengers of the *İzmir* ship were also Musa and Mustafa Refik from the newspaper *Tercuman-ı Hakikat*, A. Zihni and I. Hakkı from the newspaper *İkdam*, Ahmed Rasim and Nazmi from *Sabah* and Ohannes Ferid from the newspaper *Manzume-i Efkar* published in Armenian.[157] High-ranking officials greeted them and after staying in Beirut for four days they departed for Damascus by an exclusive train on August 28. Along the route they were hosted by important Ottoman bureaucrats such as Nazım Pasha, who was the governor of Syria, Marshal Kazım Pasha, the construction Minister of the Hejaz railway, and by Amirul Hajj Abd al-Rahman, and a feast was held in Damascus at the house of the governor of Damascus. The delegation arrived in Maan on August 31. Apart from the high officials, a crowded group of people and the Bedouin tribes of the region also participated in the inauguration of Maan station. The Bedouins added flavor to the celebrations with their horse-riding shows. Turan Pasha granted medals to people who had contributed to the construction of the line. On one side of the medal was a picture of the locomotive and the sultan's signature and on the other side was written, "In memory of the inauguration ceremony of the Maan station of the Hamidiye–Hejaz railway."[158] European supervisors who were at the ceremony found the Hejaz railway commendable in terms of technique and organization.[159] A person named Abdullah who was sent by the Amir and Governor of Mecca had sent a telegraph of praises to Izzat Pasha on September 26, 1904.[160]

Authors who attended the ceremony published important books and articles on the Hejaz railway. We have often referred to the writings of Mygind and Auler on the Hejaz railway and this fact must have attracted attention. Auler's two books on this subject contain especially important information. On the request of Abdülhamid, Auler attributed both of the works to the information he obtained during his two trips to the Hejaz railway region. In his impressions of Maan, Auler also made an interesting comment, saying that the positive effects of Wilhelm II's trip to Damascus in 1898 were still continuing.[161] Mygind, who was another guest of Abdülhamid, was the author of a work titled *Vom Bosporus zum Sinai*. After attending the opening ceremony as a guest, he started writing a second book, *Syrien und die türkische Mekkapilgerbahn*, in which he spoke highly of the ceremony.[162] Economy writer Musa, whose article series on the Hejaz railway we have mentioned in the third chapter of the present work, emphasized the importance of the Hejaz railway in his article on September 20 and pointed out that the journey between Damascus and Maan, which had taken 11 days by camel, would from now on last 28 hours.[163] One of the bureaucrats who was among the Maan delegation (probably the member of Military Review Commission Mehmet Raci) wrote a document (*varaka*) that stated his desire to create a work of art about the Hejaz railway on the basis of the data he had obtained from Izzat Pasha and Meissner Pasha and expressed his sincere thanks to both of them.[164]

Thus, a 621 km stretch of the Hejaz railway, 460 km of which was the Damascus–Maan line and 161 km the Haifa–Dera line, was opened to passenger and freight transportation. Trains between Damascus–Maan had started to operate at 30 km per hour.[165] Once Maan Station was opened, the construction gained momentum.[166] Consequently, apart from the Hejaz Railway Construction Ministry, which conducted the construction work, establishment of another administration that would oversee the operation of the railway had become necessary. This way the construction and operation of the railway would be undertaken by two separate administrations.[167] Monsieur Guedent, a French citizen, was appointed as the general director of the Hejaz Railway Operations Administration. Apparently too much was demanded of him at the beginning and he could not meet the expectations.[168] And in 1908, because of complaints coming from Hejaz railway employees about his unkind treatment and because the Ottoman administration was dissatisfied with his performance, he was replaced by Haji Muhtar Beg.[169] Also, when the railway had started to proceed southwards, the construction management had also moved from Damascus to Dera in order to manage the construction more easily.[170]

In 1906 the railway reached Tabuk, located 233 km south of Maan, and as always it was opened with a ceremony held on September 1. A delegation was present at the inauguration ceremony upon notice of Osman Ferid Pasha, the commander of Medina. This delegation had set off from Medina and been greeted by the Minister of Construction Marshal Kazim Pasha. According to the information given by Rashid Abdul Quadir, who was among the delegation, the settlement from Tabuk to Madain Salih had been completed. Trains were shuttling between Maan and Tabuk. On the way back, they reached Maan after ten hours travel. Although it had previously taken 17 days to go to Damascus from Tabuk, the commission members completed their journey "as a comfortable train travel that lasted for 30 hours". According to Rashid Beg "the Bedouin residents of Tabuk, a city that was gradually becoming prosperous, had started to be occupied with art and agriculture thanks to the sultan". A mosque located in the abandoned region had been rejuvenated and the construction of a hospital had commenced, sacrifices and prayers had been conducted on the accession anniversary of the sultan.[171] Besides, with reference to the opening of the line, 372 nickel medals had been stamped for those who had devotedly worked throughout the process, 23 of these medals had been granted to urban sheiks, public servants and officers, and 35 medals had been granted to the operators and dispatchers.[172]

The following year the railway works gained momentum and 287 km of new railway track was installed, reaching al-Ula. The ceremonies held on September 1, 1907, on account of the opening of the Tabuk–al-Ula line once again turned into a propaganda feast, with the participation of senior officials, public and the media. Upon the telegraph of Muhtar Beg announcing that the railway was moving forward towards the Hejaz region, people prayed for the sultan and made preparations to welcome the *Hatt-ı Ali* (Sublime Line), which had crossed the Hejaz borders.[173] Auler Pasha joined the trip on Abdülhamid's request and issued his second work on the Hejaz railway, which was a follow-up to the brochure he had written in 1906.[174] The railway's arrival at al-Ula and the inauguration ceremony were included in this latest work.[175]

From this date onwards, the tone of the reports by foreign diplomats serving in the Ottoman Empire changed completely. Doubts concerning the feasibility of the Hejaz railway had been replaced by positive expectations of the completion of the railway. For instance, Izzat Pasha was dubbed the mastermind of the project in the report by the British Embassy in 1906, saying: "The success of this project has probably surpassed his own expectations."[176] Also the report of Ambassador Nicolas O'Conor, dated 1907, mentions that the return of the Hejaz railway to Abdülhamid had reached inestimable levels:

> The astute policy which induced the sultan to pose before 300,000,000 of Muhammedans as the caliph and spiritual head of his religion, and in bringing home to his subjects the fervour and energy of his religious feelings by the construction of the Hedjaz Railway, which, in the near future, will afford facilities to every Moslem to perform the pilgrimage to the holy places of Mekka and Medina.[177]

The Austrian Ambassador Pallavicini took a similar point of view as he was reporting the developments

"... The modern world had defined Abdülhamid's statements of constructing the Mecca line without any foreign capital as an utopia that was impossible to realize. ... But as time passed, everyone watched in amazement as all the financial and technical difficulties were overcomed one by one ... to such an extent that when one consideres all the progress made by 1908, it is without doubt that the railway will be completed as planned."[178]

In one year, Muhtar Beg and his engineering staff of Ottoman origin had successfully completed the 323 km section between al-Ula and Medina, the area where non-muslims could not enter. This showed that the Ottoman engineers had gained enough experience in constructing railways. It is understood that while the construction proceeded towards Medina, work was also carried on from Medina to the north. According to the telegraph of the Medina Guardianship, " ... all the officials and military personnel, thousands of people and hajjis had gotten ready and started excavation works from the part called the Hamidiye Gate. Prayers were said for the sultan".[179] And according to another telegraph by the guardian of Medina, "people of Medina, children, adults, old and young people, everyone devotedly worked in the construction by singing songs, reading poems and holding their flags in their hands".[180] It is understood that to be able to organize the inauguration ceremony on the accession anniversary of Abdülhamid, the construction of bridges and similar infrastructure had been temporarily approved.[181]

In the end, the ceremony was held on time, on September 1, 1908, and naturally maximum effort was expended for it to be magnificent —the whole city and the station had been illuminated by electricity. During the ceremony residents of Medina carried Haji Muhtar Beg, the engineer who constructed the last part of the line, and Kazım Pasha, the director of the celebrations, on their shoulders.[182] The Muslim media had announced the inauguration ceremony of the Medina station to the Islamic world. Many journalists were hosted at the ceremony and once again their expenses were paid by the Ottoman government. In order for this event to be heard by most people, journalists were permitted to use the telegraph lines free of charge. Tevfik Beg from the newspaper *Sabah* and Ali Yusuf from the *Times* were among the guests.[183] The Ottoman Empire also expressed its thanks to the editor of the *Times* newspaper for the friendly approach in the article published on August 28 about the Hejaz railway.[184]

However it is true to say that the insurgence of the Young Turks had overshadowed the celebrations. After the establishment of constitutional monarchy, Abdülhamid's powers had been restricted; Izzat Pasha, the mastermind of the Hejaz railway, had been obliged to flee the country and his name was not mentioned at the ceremonies. All this created doubts about the pro-Islamic policy of the Ottoman Empire.[185] But still, as the new regime had embraced the Hejaz railway and paid attention to the celebrations, the opening had been full of excitement and ended with the hopeful crying of the crowd: "on to Mecca!"[186]

The opening of the Medina station had been greeted with enthusiasm by the Muslims in some Islamic countries. For instance, the *Times of India* had published accounts of the arrival of the line at Medina, commenting that this success revealed the solidarity of Muslims and the potential power of the Ottoman state.[187] Again, the Islamic sect leaders in Indian cities such as Mumbai, Calcutta and Yangon had announced that the opening of the Hejaz railway would be celebrated in mosques with solidarity. The text that they prepared emphasized the importance of September 1 for the Islamic world and stated that the caliph with his loyal followers had revealed the "spiritual power" of Islam. After prayers were said, the text was read in mosques.[188]

G) EXTRAORDINARY PROBLEMS

1- *Water*

After 1906 the construction of the lines gained great speed. After Maan, the Hejaz railway proceeded across a plain where natural obstacles did not exist. Thus, there was no need for infrastructure like bridges, tunnels or viaducts.[189] But apart from the geographical advantages, the region also had extraordinary issues that needed to be solved. The most important problem was water supply during the construction. The only stations along the Hejaz railway that had water reserves were Damascus, Dera, Amman, Maan, al-Hasa and Mudawwara. There were many wells and reservoirs left over from Roman times in the north of Maan, but the reservoirs were shallow which led to evaporation of water at high temperatures.[190]

Besides, the desert heat, which could reach 50–60°C in the shade, augmented the water requirement beyond normal levels. The trouble was that the natural water resources completely disappeared as one moved towards the southern regions, where water demand was at its highest. For instance, the 113 km region located between Maan and Mudawwara and also the 190 km stretch between al-Akhdar and Madain Salih were completely lacking in water.[191]

The Hejaz Railway Administration had tailor-made water tanks. Although it also tried to restore the wells and reservoirs in the region, this appeared to be more laborious than expected.[192] So the administration made attempts to drill new artesian wells.[193] In some places before Medina, especially al-Akhdar and Madain Salih that were located in the southern part of the region, the wells were dug 110 meters deep. When these efforts did not yield result, it was decided to transport water by wagon. Tank wagons designed by Süddeutsche Waggonfabrik could store 8–12 m^3 of water. The subsequent increase in operational costs was not hard to predict.[194]

Another method which was highly costly and troublesome was the transportation of water by camel. Given the fact that it took three days for a camel caravan loaded with water to travel 100–150 km, only a minimum part of the demand could be satisfied in this way. Both methods (transporting by wagon or camel) were very costly, but were nonetheless implemented. Water was transported by wagon up to the points where rails were installed, and from these points water was transferred to camels and was transported to construction sites. For instance, on July 18, 1904, in order to send water to the region, it was decided to immediately allocate 50 camels and 100 extra removal wagons in addition to the 160 removal wagons already available.[195]

As the regions where construction continued were uninhabited, the function of camel transportation was not only transferring water. All the needs of the workers, especially food, and railway materials were transported by camel. However, the Ottoman administrators expressed the difficulties that arose. For example, in 1907 a report by the Meclis-i Vükela stated that the camel owners were hesitant to come to the railway without a guardian.[196]

Renting camels increased the costs to such an extent that the Hejaz Railway Administration decided to buy 400 camels in order to solve the water transportation problem. But camel transportation and care was only possible for those experienced in the Bedouin lifestyle. When the Ottoman soldiers could not manage to do this, many of the camels died. When camels were rented, the camel owners were expected to provide a new camel in case a camel died along the journey.[197] The Bedouins who made a living by the rental incomes of their camels were stealing the camels that wandered away from the rails. As a result, the business of renting a new camel for one mejidie per day was delegated to a contractor from Damascus engaged in the caravan trade. In addition, they continued to rent camels from the Bedouin tribes in the vicinity. But when they saw that the cost of only one camel for three months was 90 mejidie they again decided to solve the transportation problem by buying camels. This time professional camel keepers were employed to avoid camel deaths. For instance in 1905 the Komisyon-ı Ali decided to appoint an expert in the treatment and care of camels at a salary of 250 kurush.[198]

Providing the desired number of camels of the desired quality was one of the most important problems that the caravan directorates encountered. The Grand Vizier of Sultan Murat III, Ibrahim Pasha, had established a foundation in 1586 in order to solve the problem of providing camels, and donated 600 camels to this foundation. It is understood that the foundation failed because of the resistance of communities that traditionally rented camels.[199]

As it was not possible to transport water by camel to distant places, water stations were built in intermediate areas. In the intermediate water stations, barrels were buried underground. Water trains would fill these barrels periodically and the camel caravans would obtain the water, which they would then transport to distant places. All these efforts could solve the problem only to a certain extent. The water supply remained an important issue for the regions located to the south of Maan.[200] The scope of the water problem was such that even Kaiser Wilhelm was compelled to look for a solution to it. The Kaiser had stated to one of the Ottoman governors that the Ottoman government could make use of a German expert who could say whether there was water in a region or not by using a special tool.[201] Ottoman engineer Mustafa Şevki Atayman, in his memoirs, talked about a German who claimed to know where to find water alongside the line by a tool that he had invented.[202] This man was most likely the one the Kaiser had in mind.

The importance of water in the hajj organization can be understood from the fact that two chief water-bearers were appointed from the Enderun to supply the water needs of the hajjis starting from Scutari. The administrators on the way were ordered to oversee these two water-bearers.[203]

Interestingly enough, while some parts of the region experienced a grave water shortage, floods were damaging the railway in others. Ottoman engineer Mustafa Şevki Atayman reported that once heavy rain swept away the 18 stone bridges on the 40 km line which "happened to be on the Damascus direction of Maan". In order to be protected against floods, 799 headwalls, 462 bridges and 271 viaducts had been constructed only between the region between Damascus and Mudawwara.[204]

Another difficulty that the Hejaz Railway Administration encountered due to the characteristics of the region was sandstorms.[205] During his expedition, Muhtar Pasha had stated that the sandstorms would damage the lines after Mudawwara. In order to protect the rails from sand, walls were constructed from clay and stones, which were abundant in the region.[206]

2 – Fuel

Supplying fuel materials was another important challenge. Syria and Arabia were poor in terms of forests and coal beds. So coal, which was used as locomotive fuel, was imported from other countries and stored in Haifa and Damascus. Transportation of the imported coal to the Hejaz region was also increasing the costs. We know that this was the reason for having the special wagons designed by Süddeutsche Waggonfabrik.

Supplying coal was getting more difficult as the rails progressed along the 1,400 km line. In the region where in winter the temperature was below zero, resolving the fuel problem was of vital importance both to generate the energy needed for the trains to operate, and to heat the station buildings and tents. Besides, the problems of lighting and cooking for thousands of people working in the construction made it obligatory to import coal. In order to ease the problem, the Eregli coal bed, which operated under the Ministry of Naval Forces, "was endowed to the Hejaz Railway Administration upon the sultan's orders".[207] A Komisyon-ı Ali report dated October 23, 1906, stated that 25,000–30,000 liras were required to supply annual coal demand. So it was decided that the Eregli coal beds, which needed a minimum capital of 70,000–80,000 liras in order to operate, were to be lent to Emin Effendi without any capital injection.[208]

In 1906 Auler Pasha suggested an alternative solution to the fuel problem. Auler observed that the railways being constructed in the Karakum region of Russia had the same problems and they had resolved these problems by exploiting existing oil reserves in the region. The oil from Mosul could possibly solve the fuel problem of the Hejaz railway as well.[209] For Auler's suggestion to be brought to bear, firstly the Bulgurlu–Mosul part of the Baghdad railway and the Damascus–Aleppo part of the Hejaz railway needed to be constructed and these two lines had to be connected at Aleppo. So in the short run the problems would be solved by imported coal.

The Administration also had to import the lime to be used in the preparation of the plaster. Catering food to thousands of construction workers was also another problem, as the region was scarcely settled. Overcoming these difficulties was getting harder and harder as the construction progressed through the Hejaz deserts towards Medina.

3- *Labor*

The most problematic issue in the Hejaz railway construction seemed to be finding workers, master-builders, skilled laborers and other employees to work in the construction. The labor problem in Maan was resolved to some extent by recruiting settlers as wage laborers.[210] Besides, the Muslim people who lived near the railway route were forced to work by law. This forced labor was generally used in excavations. But as the rails moved towards the south—to the desert region—cities, towns and villages disappeared. Employing people by force in order to work in distant deserts with no security would also create inefficiency, a problem which the Hejaz Railway Administration would not be able to surmount.

Therefore a new method that would provide financial and social benefits to workers had to be applied in order to find a voluntary workforce. Such a solution was found and successfully implemented. Mainly railway troops composed of soldiers were used in the construction. The soldiers who worked in the construction of the railway were exempt from one-third of their military service. In addition, the government payed 1 kurush to the soldiers for the excavation of each cubic meter of land.[211] Consequently the soldiers gained the opportunity to contribute to the construction of the "sacred line" and at the same time be discharged earlier and save a little bit of money. However, Seraskier Rıza reported that, "the soldiers would desert sometimes and desertions from the other military troops were also common. Subsequent notices from the military had been sent to the relevant provinces for the runaway soldiers to be immediately captured and sent to the railway lines."[212] This shows us that the method was not working as smoothly as expected. Thus, when 3,000 soldiers were requested from the Sixth Army, it was asked to recruit those who volunteered to work and not force anyone.[213]

After announcing his decree on the Hejaz railway, Abdülhamid had commanded the establishment of a railway battalion using some troops of the Fifth Army. Soldiers who would be beneficial for the railway construction, craftsmen for example, would be chosen for this battalion. Military engineering officers were recruited to the command echelons of the railway battalion. Immediately after the first railway battalion, a second one was established.[214]

In the beginning the construction proceeded slowly, but as the soldiers gained experience it gained momentum and they started to install 2–3 km of rail every day.[215] Soldiers who worked in the railway construction stayed in tents set up to the right and left of the railway[216] Each battalion had its own mobile kitchen and oven where they cooked and baked bread. In winter the soldiers worked in uniforms made of baize prepared by the Ministry of War and in summer they worked in white garments known as *kefiye* which to some extent protected them from sun.

When this practice proved successful, new railway battalions were established, each one consisting of 1,000 soldiers. The telegraph platoon of the Fifth Army, which had 50 personnel, was also transferred to the Hejaz railway. The railway battalions made up of 5,650 soldiers of the Fifth Army based in Syria had started to work on the Hejaz railway in the first two years of the construction.[217] A military commission convened on February 14, 1902, decided that the operation battalions working in the construction would be employed every six months and in rotation, and a pharmacist, a physician and a surgeon would be assigned to each battalion. When the construction reached Maan, the number of worker-soldiers had reached 7,300. Among them were 200 mariners and 100 soldiers from the Acre Artillery. The duty of the aforementioned 300 soldiers was to carry out the loading–unloading operations of imported railway materials which were transported to Haifa by ship. As mentioned above, a port had not been constructed in Haifa, which was why loading and unloading operations had to be carried out carefully.[218]

In the beginning, soldiers were charged with ordinary tasks such as excavation and collecting stones. Later, they began to install the rails and traverses too. Craftsmen such as carpenters, smiths and bricklayers were recruited from among the soldiers.[219] When the construction approached Medina, the majority of the soldiers had fallen ill and, although employment of local people in the excavation works has been considered as an alternative solution, this idea was abandoned due to the "ungentle character" of the locals. Instead two troops out of the seven that were located in Medina were deployed.[220]

a) The Organization of the Railway Battalions

In the Hejaz railway construction, three separate railway battalions served in three different regions. The survey troop that served in the front region would determine the route across which the lines would pass, assessments were carried out in the middle region and the troop in the back would continue the construction activity. As the survey troop moved forward, the assessment troop would take its place, and when the measuring troop moved forward, the rails and traverses would be installed in the area as the necessary surveys and measurements were completed.

The survey group was supposed to confirm the route and report it to the Railway Administration. A survey troop was generally formed under the leadership of a railway engineer and included two qualified engineers, one pharmacist, ten soldiers and 20 cavalrymen. There would be camels, horses and other pack animals to carry the necessary materials. A similar organization was necessary for the field assessment troop. Their most important duty was to prepare the altitude map of the surveyed field. Then the construction troop would deal with excavation and grading of the ground, cleavage, ballast preparation and installing rails and traverses on areas where the survey and measurement works had been completed.[221]

As the infrastructure work required qualification at least to some extent, Ottoman soldiers had been subjected to an accelerated training. For instance, in order to provide a fast training to the fourth company soldiers of the number one operation troop in areas like masonry and carpentry, very high wages were paid to some of the teachers.[222] As a result, railway battalions obtained the knowledge and experience needed to conduct any kind of construction work.[223]

In a nutshell, despite its drawbacks, deploying soldiers in the construction process proved successful. When the advisor engineer Gültstein stated, "the biggest contribution to the progress of this railway comes from the Ottoman soldiers. Without the work of the disciplined Ottoman soldiers, which are also in the required quantity, the line wouldn't be realized". he was not wrong at all. Gültstein also accentuated the fact that coordinating the officers who worked as supervisors and civilian engineers was actually not easy: "because Bulgaria and Serbia had made an attempt to construct a railway using soldiers and failed".[224] Recruiting soldiers in railway construction was a method that the Russians used in their Transcaucasian railways.[225]

b) Benefits Provided to the Workers

Extra money was paid to the soldiers of the operation battalion in proportion to the work they did. For example, one kurush was paid for grading one cubic meter of land and placing one cubic meter of stone on the railway. The typical work day was ten hours. It would begin before sunrise, at noon there would be a break of two hours, and then work would continue until sunset. Given that a soldier could grade three cubic meters of land in one day of hard work, he could only earn 3 kurush. Moreover, the soldiers could only work five days a week as Fridays were leisure days and Thursdays had been set aside for cleaning.[226] In that case it was pretty difficult for a soldier to earn a decent additional income.

Apart from the soldiers, a side income was also provided to officers who served in the railway troops. The most important advantage the officers obtained by working in the Hejaz railway was receiving a special promotion after two years of service. This method had been in use in the Ottoman Empire for people who worked in arduous conditions.[227]

The term of service in the railway battalions was reduced from three to two years due to health problems arising from difficult working conditions. But still there were many people who caught diseases like cholera, anemia, scurvy, and dysentery because of the challenging desert conditions, malnutrition and inadequate supply of water.[228]

After 1904 health conditions partly recovered. Several hospitals had been opened along the railway. The railway administration tried to elevate the level of health services by increasing the number of medical personnel serving in the railway battalions. In 1902, the military commission had decided to employ a pharmacist, physician and surgeon in each of the battalions.[229]

Legal regulations that provided health insurance to the Hejaz railway workers were in place. According to the bylaw issued in 1904, there would be a deduction of 1 per cent from the salaries of all officials and employees working in the Hejaz railway in return for free medical treatment. If deemed necessary, the employees would be transferred to hospitals in the nearest region, but their families would only benefit from the treatment in their place of residence. On the other hand, those whose illnesses were not diagnosed by the surgeons and who left their work without permission would not be paid their wages and salaries. The clauses of this bylaw were valid for officials and workers who got sick or were accidentally injured and became disabled at work. Article eight of the bylaw concluded that treatment costs would be taken from those who contracted syphilis and similar diseases and those who got sick or injured due to non-work-related factors, and these people, too, would not receive their salaries and wages during their treatment process.[230]

In case of injury or death, salaries would be paid to the workers or their families. For instance, sappers Dervish and Hayrullah, who had gone blind during their service, would receive salary regularly. The family of Mehmed, who had been killed by the Bedouins in the Hediyye section, received a certain remuneration. Apart from these health and social relief arrangements, officials and employees who worked in the railway as permanent personnel were entitled to a pension.[231]

CHAPTER 5

FORCES RESISTING THE HEJAZ RAILWAY

A) BRITISH OBSTRUCTION OF THE AQABA LINE: THE TABAH CRISIS

1- *Significance of the Aqaba Line*

Since the idea of building the Hejaz railway was first formulated, emphasis had been put on the advantages of building a branch line to the Gulf of Aqaba, situated on the coast of the Red Sea.[1] With the progress of the Hejaz railway towards Medina, the strategic and economic advantages that would be conferred by the Aqaba line gained more popularity and it became an indispensable part of the agenda. Auler Pasha, for example, is reported to have mentioned the problem of supplying railway equipment and other required material to the south of Hejaz railway construction. He further claimed that the most rational method that might be used to solve this problem would be to build a connecting line from Maan to the Red Sea.[2]

We have already mentioned that the Haifa–Dera railway had been serving that purpose. However, shipping goods from Haifa through Maan, which was 400 km away, and carrying the shipment further ahead, was not rational at all. Therefore, the cheapest way to ship goods to the southern part of Maan would be to build a new branch to the Hejaz railway. The Gulf of Aqaba stood out as an ideal location.[3]

The Aqaba line was important not only in economic terms but also because of its strategic advantages. Movement of goods and soldiers was done via the British-controlled canal and this posed a major threat to the Ottoman Empire. The British might close the canal in case of war or for any other reason, which would sever connection to the region. Constructing a branch line from Maan to Aqaba would help the Ottomans dispatch goods and soldiers much more quickly without using the canal. For example, troops in Syria could be transferred to the Red Sea via the Hejaz railway. From there on, sea transportation would eliminate possible chaos in the Hejaz and Yemen and spare soldiers a toilsome march from Maan to the Red Sea coast under the dazzling sun.[4]

Ottoman authorities had already started considering an alternative line as a result of a series of incidents. In 1904, for instance, a total of 1,000 Ottoman soldiers were supposed to be sent from Yemen to Aqaba by ship, then walk 80–100 km to Maan, reach the Hejaz line again and board the train. However, the soldiers would have to walk through the desert and they rebelled against this. The captain had no alternative but to turn the route towards the Suez Canal, which, in the end, made it necessary to pay an entrance fee to the British. The Ottoman Commissar responsible for Egypt, Ahmet Muhtar Pasha, took advantage of this incident to revive the Aqaba line project. According to the Pasha "recent incidents indicated how critical it is to build the railway between Aqaba and Maan".[5]

There were indeed rebellions among Ottoman soldiers, particularly among those who were dispatched from Damascus to Yemen and revolted against marching from Maan to Aqaba. One of the greatest problems encountered was that of water supply in the desert. Water in the oases close to the resting spots was strictly controlled by Bedouin tribes. Therefore, being on good terms with them was *sine qua non* for ensuring water supply. For example, in 1905, in order to provide water for 15,000 soldiers, 1,500 camels were hired from the Huwaytat tribe.[6]

Furthermore, according to statistics kept at the time, Ottoman authorities would not have to incur any extra expenditure to build the Maan–Aqaba railway. Ahmet Muhtar Pasha claimed that the transit fee that they would otherwise have to pay to the Suzel Canal Company was high enough to cover the cost associated with this line. Thanks to this approximately 120 km railway, the total cost of transportation would be reduced, not to mention other advantages such as the feasibility of this new route. The members of the Council of Ministers were unanimous about of Maan–Aqaba railway: this railway would be utterly "beneficial without an iota of doubt", as Grand Vizier Ferid conveyed to Abdülhamid.[7] Finally in 1905, Abdülhamid decided to launch the construction of a branch line from Maan to Aqaba.

The above-mentioned line would clearly boost the likelihood of success in case of military operation against Egypt. The odds were that the British authorities would oppose the Aqaba railway project for military and political reasons. In the meantime, radical Islamist newspapers fervently stressed the military and strategic advantages associated with the line, which, most probably, led to stronger British resistance. Originally, the construction was supposed to start in April 1905;[8] however, Abdülhamid had to postpone it. Although he had accentuated the financial factors, British influence in the region was the main reason behind the postponement.[9] Despite all the efforts, the Tabah Crisis of 1906, which led to a major conflict between Ottoman and British forces, could not be averted.

2- *The Tabah Crisis*

Once the Ottomans began to build a telegraph line between Maan and Aqaba, and increased the number of the soldiers to 2,000,[10] tensions with the British authorities in Egypt inevitably escalated. One of the measures the British government adopted in the northeastern part of the Sinai Peninsula was deploying some troops to Tabah, located 12 km from Aqaba. However, when the British Major Bramly arrived in the region, he realized that the Ottomans had already made that very move. By January 1906, an Ottoman troop leaving Aqaba had already established headquarters in this small village at a highly strategic location.[11] Soldiers and their arms were sent via railway to Maan, and then beasts of burden were used to reach Aqaba.[12]

The rift between the British and the Ottomans grew deeper and eventually the British government asked the Ottoman government to withdraw from Tabah on the grounds that the village was within the borders of Egypt. Their claim was based on the fact that the Ottoman state had transferred authority over the Sinai Peninsula to the Egyptian government in 1892. Accordingly, a line was to be drawn between Rafah and Aqaba and what remained to the west of this line would be governed by Egypt.[13] The Ottomans, however, argued that the government of the region had been transferred to the Egyptian administration temporarily and that the only reference document was the Agreement of 1841, which had delineated the Egyptian border between Syria and Arish.[14] According to Ottoman rulers, the area between Aqaba and Arish was uninhabited and was used as a transit point by Arab tribes. Consequently, Ottoman intervention had hitherto not been required. Taking advantage of this situation, the Egyptian government had had the chance to take hold of some of the coastal regions of the Sinai Peninsula.[15] In fact, the authority of the Sublime Porte, which they attempted to maintain, existed only in theory. Starting from the Mehmet Ali era, the Sinai Peninsula was practically governed by Egyptian khedives. And now, it was impossible for the British government to accept the reestablishment of Ottoman authority in the region, given the critical circumstances. The analysis of the *Morning Post*, published on May 12, 1906, reflected the widespread opinion of the British authorities:

Aqaba is 70 miles from the city of Maan in line with the Hejaz railroad. The Ottoman Army in Maan are positioned closer to the Suez Canal than the Egyptian army. Therefore, to have the strategic edge, the number of Egyptian soldiers should be increased. For this reason, the Ottoman proposal to change the borders cannot be accepted. In fact, the existence of the Ottoman army adjacent to the Egyptian borders already stands as a major threat.[16]

Let us now focus on another newspaper article that upheld the Ottomans' position. The article is significant in that it was published in *Muhammedan*, one of the periodicals (others included *The Moslem Chronicle* and *The Comrade*) published with great difficulty by Muslims in India to present an alternative to the British-backed newspapers.

> The Hejaz train will tie an ever-lasting knot between Arabia, caliphate and the Sacred Rule of Islam. As long as the British do not leave Egypt, the Ottoman State will not, and should not, consider changing the basic policy of protecting its own interest in the region ... It is the sultan's duty to maintain power around the Red Sea and use authority to control the surrounding regions. Those newspaper articles that claim that "The Ottoman statesmen is acting in obedience to the German Emperor," are nothing but totally wrong and irrelevant ....[17]

The Tabah Crisis affected the bilateral relations to a greater extent until mid-1906. The British government kept on demanding that the Ottoman troops withdraw, while the Ottoman Commissar for Egypt, Ahmet Muhtar Pasha, rejected this demand, claiming that Tabah was actually a village bound to Aqaba. The proposal by the Egyptian governor to convene a joint commission was rejected by Ottoman politicians. After all their demands had been rejected, the British sent the battleship *Diana* to the Gulf of Aqaba as a threat.[18] However, the Ottoman government kept claiming that Egypt was legally a part of the Ottoman Empire. In the meantime, two Ottoman army officers went to Egypt to interview Ahmet Muhtar Pasha and returned without having made official contact with the British and Egyptian officers.[19] Abdülhamid gave full authority to Muhtar Pasha during the meetings. Both the British garrison in Egypt and Ottoman forces in Aqaba were provided with support.[20]

When finally the British government realized that they would not be able to dismiss Ottoman forces by simply threatening, they gave an ultimatum to the Ottoman State via Ambassador O'Conor.[21] Dated May 3, 1906, the ultimatum made it clear that the situation would deteriorate should "the Ottoman forces not withdraw from Tabah in ten days".[22] While Britain strengthened its military hold in

Egypt, they also stepped up the pressure by sending warships from Malta to the Aegean coast.[23] Subsequently, the Ottoman state announced that they would satisfy British demands in the middle of May. In other words, the Ottomans agreed to withdraw from Tabah and let a joint commission resolve the issue. After heated debates in the commission, an agreement was finally signed on October 1, 1906, which finally drew the border between the Ottoman state and Egypt according to British demands. "The Ottomans bluffed, but bluffed in vain."[24] Seven years later, though, Fitzmaurice also used the same wording: "I took no notice of such manoeuvres. Knowing that it was mere bluff... the ex-sultan used Panislamism to frighten us both here and in Cairo. It was ignored and no unpleasant outcome ensued."[25]

In contrast to what some of the British newspapers had claimed, British obstruction of the Aqaba line was not based on economic rivalry with the Suez Canal. In fact, the railway connection between the region and Europe would be by all means longer than the Suez Canal.[26] As already stressed by the *Daily Telegraph*, the real reason behind this obstruction was more strategic than economic. In another article, Aqaba and Kuwait were referred to as two major strategic spots for the British. In a similar way, Sir Edward Grey, who made reference to these two places, claimed that the line between Maan and Aqaba would help Ottoman soldiers reach the Red Sea, which posed a threat for the Egypt–Red Sea–India road.[27]

The newspaper article mentioned above was presented to the Prime Ministry by Germany's ambassador in London, and the Hejaz Railroad report issued by Auler Pasha was translated into English and sent to Britain. This demonstrates the political significance of the issue. The fact that the Germans proceeded towards the Persian Gulf via the Baghdad railway over Mesopotamia created further tension. The British believed that the Aqaba line would give the German–Ottoman alliance leverage to threaten the Suez Canal.

In fact, the idea that the Germans were endeavoring to gain influence in the region through their relationship with the Ottoman state was not imaginary at all. The diplomatic correspondence also reveals that German Emperor Wilhelm II observed the Aqaba/Tabah crisis intently. Reading a report by Marschall, he wrote a note in the margin of the report saying that he gave no credit to the British for what they claimed. He believed that British tactics, aiming at breaking Ottoman resistance, were "cheap".[28] Austrians also supported the Hejaz line. Schick, for instance, felt that France had lost power in the region while the Austrians together with the Germans grew stronger.[29] Grand Vizier Ferid Pasha informed Abdülhamid that France would probably demand to take over the management of the Aqaba line for a period of nine years,[30] which proved that the imperialistic relations were rather complicated at that time.

The importance of the Hejaz railway for the Germans to gain power in the Middle East has already been explained in detail in the chapter on the Haifa branch line. Thanks to the Baghdad and Hejaz railways, Germans believed they could outflank the British in India and Egypt in alliance with the Ottomans. Paul Rohrbach underlined the importance of the project for Germany as follows:

> Britain ... has one Achilles' heel. That is Egypt. ... The conquest of Egypt by a Muslim country like Turkey will endanger Great Britain's influence with sixty million Muslims living in India. ... However, for Turkey, reclaiming Egypt prior to the completion of a decent railway network in Asia Minor and Syria would be an impossible dream".[31]

At this point, we should refer to Ambassador Marschall's remarks: "I know for certain that the possibility of re-conquering Egypt is in the minds of many serious Turkish statesmen, especially since the Hejaz railway was built".[32]

Baron Max von Oppenheim wrote a related report. A renowned archeologist, Oppenheim was knowledgeable about Islam and the Arabs, and was the acting German consul general in Cairo. He was influential in formulating Germany's Middle East policies, particularly as regards subversion and the incitement of a "holy war" against the Entente powers.[33] According to him, a possible connection of all the railways across the Persian Gulf, Hejaz, Yemen and İstanbul would ruin British attempts to sever the holy cities from the Ottoman Empire and to reduce the power of Ottoman caliphate in the Islamic world. "If the Ottoman State and the caliphate reinforced their authority in the region, the British authorities would have to withdraw from Egypt. In impede this, Britain would aim at taking over Mesopotamia and the Persian Gulf along the Baghdad railway. Therefore, Germany needed to beware of an imminent British assault".[34]

In another report, Oppenheim explained that Britain had already been trying to sever the holy cities from the Ottomans and the caliphate. With this aim in mind, they endeavored to invade Mecca and Medina over Yemen in the south, Aqaba on the northwest and Kuwait from the northeast. These were the typical hajj routes. The Ottoman military in Yemen had lost blood because of British provocation. Oppenheim stated that "When the Hejaz railway reaches Medina next year, it will stand as a shield against the threat coming from the northeast. Right now, they are trying to focus marching from the northeast." He also stressed that the unrest in Muslim colonies was an indication that Islam was still alive. Britain's biggest fear was Muslim colonies' revolting under the leadership of the caliph and containing Great Britain. Oppenheim stated that a "less cowardly" Ottoman sultan could well have turned this latent power in the Islamic world against the British. As long as this potential threat persisted, England would have to avoid a possible war in Europe, especially with Germany. The German emperor had to be on good terms with the Ottoman sultan, while Britain was also aware of this fact and tried to manipulate the sultan by pursuing a carrot and stick policy.[35] All in all, Oppehheim clearly stated that the caliph/sultan should be manipulated to pursue a pan-Islamist policy. However, German authorities still tried hard to create a friendly impression in their messages to the British. For instance, Prime Minister Bülow said to the British ambassador that they were in favour of an amicable solution to the problem and that they had advised the sultan to satisfy British demands.[36]

Finally, let us try to interpret the inconsistencies in the Ottomans' attitude during the Tabah crisis. Despite reacting radically at first, the Ottoman government later accepted all British demands, and even went so far as to state that Tabah was invaded by the Ottoman commander in charge solely upon his own initiative. As already mentioned above, the radicalism in the Ottoman policies in the Tabah crisis derived from one main actor on the scene: Egyptian Extraordinary Commissar Ahmet Muhtar Pasha, who rejected all the demands posed by the British during the process. In 1897, Ahmet Muhtar Pasha had already offered Abdülhamid to connect Damascus and the Suez Canal by a railway. The interesting point is that, in this very correspondence, British obstruction had been predicted. Muhtar Pasha maintained these ideas in the years to come,[37] and when he finally had the chance to put them into practice in 1906, he did not hesitate for a minute. Lord Cromer also reports Muhtar Pasha's words telling him that the Ottomans had to be present in Tabah in order to ensure the construction of the railway.

This leads us to conclude that the Ottoman commander in Aqaba actually invaded Tabah upon his own initiative and this impulsive move was endorsed by the Egyptian extraordinary commissar. Since the Ottoman ruler already desired to build the Aqaba line, such a move sounds logical. However, when the British government issued an official ultimatum and threatened Abdülhamid by sending the Mediterranean naval forces to the Dardanelles and the battleship *Diana* to Aqaba, the sultan had to back down.

In this respect, Abdülhamid's chief clerk, Tahsin Pasha, remarked that the sultan always tried to avoid problems with the British government and therefore was rather concerned about the Aqaba crisis. Abdülhamid was cautious: he did not want to provoke the British for fear that they could try and take advantage of the crisis to block the railway construction altogether.[38] In reality, this fear was a bit far-fetched. The major objective of the British, apart from blocking the Aqaba line, was "to legalise their invasion in Egypt, which they achieved by 'forcing' the sultan to draw the line between Egypt and the Ottoman Empire".[39]

Although the crisis had ended in favour of the British, the British kept sabotaging the Hejaz railway. For instance, on the grounds that contagious illnesses spread during hajj, the British asked the Ottomans to authorize an international health supervision system in Mecca and Medina. Imposing some daunting measures, such as long hours of quarantine, they tried to deter the Muslims within their colonies from going on the hajj. Hajjis who reached Beirut over Damascus via the Hejaz railway were unnecessarily kept waiting in the city. The German consulate in Beirut reported that 500 Egyptian hajjis were kept waiting for three weeks. The consulate remarked that this attitude of the British not only meant unfair treatment of Muslim pilgrims, but also a hostility towards the Hejaz railway.[40] Ottoman authorities sought ways to help Ottoman hajjis reach the Hejaz coast without passing through Egypt.[41]

On the other hand, the Ottoman authorities tried to avoid any controversy with the great powers over the issue of health. For instance, the Ministry of Health began to put more emphasis on healthcheck procedures of hajjis visiting Beirut.[42] In 1908, a new quarantine station was planned to be built on the hajj route, and a Quarantine Commission, consisting of German, English, Dutch and Ottoman delegates, was established.[43] Just before World War I, the construction of a new health office began in Jeddah.[44] Another sweeping measure was to stop the trains that departed from the health office in Tabuk, Maan, Quatrana and Amman to examine the hajjis. "Trains were not allowed by the province of Syria to proceed to Damascus until dawn."[45] The document quoted below also reveals how diligent the Ottoman authority were:

... while the hajjis returned in groups, because of the continuing plague in Jeddah, they had to be held in quarantine by the Egyptian Health Association; and afterwards, those who made their way towards the north were taken into the Tour Health Quarter for seven days; once the quarantine was over, it was decided that they be put into another quarantine for precaution either in Klazomen, Beirut or Tripoli. All the hajj boats returning to Yemen and Basra were also sent to Kamran for a ten-day quarantine; and all the Russian hajjis wishing to pass the gates of the Mediterranean and Black Sea will alsohave to go through a long quarantine process ... these decisions are taken by the Association of Healthcare and approved by the Ministry of Health".[46]

B) ATTEMPTS TO SUSTAIN THE
HEJAZ RAILWAY CONSTRUCTION

1- *Significance of the Medina–Mecca and Mecca–Jeddah Lines*

Although the Maan–Aqaba line had not been built, the Hejaz railway reached Medina in 1908. In other words, a 1,464 km section of the railway, together with the Haifa–Dera branch line, was opened. However, extending the railway to Mecca had not yet guaranteed the safety and comfort of the hajj journey. Although it was now possible to travel from Damascus to Medina by train, the hajjis still had to set out on another journey of ten or 15 days under the Bedouin threat.

Therefore, in order for the Hejaz railway to fulfill its purpose and ensure the safety of the hajj, a 74 km branch line between Mecca and Jeddah had to be built. In fact, most of the hajjis coming from India and Java travelled by sea. All hajjis reaching Jeddah by boats had to set out on an uncomfortable and dangerous journey on camels. The letter sent to the Sublime Porte by Muhammad Inshaullah in 1910 was a clear warning: unless the construction of the lines between Medina–Mecca and Jeddah–Mecca was launched immediately, it would become difficult for Indian committees to raise funds for the railway. As a matter of fact, the Ottoman government was well aware of the importance of the aforementioned lines. For instance, Hajji Muhtar Beg had written a report concerning the Hejaz railway route, explaining the Jeddah–Mecca and Medina–Mecca parts in detail and proposing two alternatives for the Medina–Mecca line. He argued in favour of the one on the west, which would follow the hajj caravan route, stating that this 74 km line would be more feasible to build.[47]

A report written in 1905 to the Komisyon-ı Ali by Kapp von Gültstein seriously suggested constructing a line between Jeddah and Mecca.[48] According to Gültstein's calculations, based on the data he had received from Muhtar Beg, the Jeddah–Mecca line could be completed on September 1, 1907, if the construction was launched immediately. The 460 km line between Mecca and Medina could then be completed by 1910. This line could be connected to the one proceeding from Damascus to the south, hence completing the Hejaz railway. If this plan could be realized, the Hejaz railway would be completed three years early, providing the Ottoman Empire an access to the Red Sea.[49]

Furthermore, Gültstein calculated all the potential profit associated with the Mecca–Jeddah line, thanks to transportation of goods and passengers. The needs of 100,000 inhabitants in Mecca were satisfied thanks to import operations conducted by the Jeddah port. The profit made from a minimum of 30,000 hajjis traveling annually would be added on top of transportation revenues, the total income reaching an unprecedented level. However, shipping the railway equipment via the port of Jeddah instead of Haifa would decrease the total cost remarkably. Gültstein also emphasized the element of confidence and prestige in the Islamic world thanks to the railway. Realizing that their contributions were not in vain, Muslims would sustain their support.[50]

An article focusing on the benefits of the line was published in *Frankfurter Zeitung*. According to the article, an average of 100–120,000 hajjis travelled to Mecca through Jeddah annually, which would increase the profit of the railway by 150,000 liras per year. The amirs of Mecca received 3 liras per person in return for ensuring their safety in the area. Since law and order would henceforth be established by the railway authority, this fee had to be transferred to their revenues—hence the calculation stated above. Furthermore, the construction would increase the number of hajjis. The author concluded that this profit could also be used for the construction of Mecca–Medina line.[51]

In the meantime, the Young Turk Revolution (1908) against Abdülhamid took place, and the leading figure behind the Hejaz railway project, Izzat Pasha, fled abroad. The official name of the railway, the Hamidiye Hejaz railway, was changed to Hejaz railway, removing the reference to Abdülhamid. However, it should be noted that the new regime of the *İttihat ve Terakki* (Committee of Union and Progress) aspired to further this significant project despite their misgivings about Abdülhamid's policies. The two lines had already been regarded as vital by Ittihadists in order to strengthen the empire against the amirs of Mecca and Bedouin tribes.[52] Unless the project was realized, the implementation of reforms planned for the Hejaz would be impossible.

Nevertheless, the political and economic circumstances set serious obstacles before the Ittihadists. The War of Tripoli and Balkans (and the financial difficulties resulting from these) constituted the main hurdle, whereas rebellions and the resistance of the amir of Mecca and Bedouin tribes were another major impediment.[53]

Ittihadist' centralist policies were protested not only by the amirs of Mecca and Bedouins but also the inhabitants of Mecca and Medina. In 1903, popular resistance prevented the collection of urban tax in Medina for street-cleaning operations. Inhabitants of Medina believed that if they failed to react, they would gradually lose their tax exemption. The electric tramway line formerly planned for Mecca was also cancelled due to the failure of the railway construction.[54] For centuries, the Hejaz populace had survived thanks to hajj donations made by the Ottoman Empire, donations by the sultan, gifts sent from the capital, and all donations granted by charitable organizations across the empire.[55] The inhabitants surmised that a stronger central authority in the region could undermine their autonomy and challenge their privileges, such as tax and military exemptions.

2- Resistance of the Bedouin

When the construction approached Medina, Bedouin tribes rallied against the Hejaz railway. Bedouins felt that once the Hejaz railway began to operate, they would lose their economic power. They made a living primarily by renting camels to hajjis and supplying them food and water. The Ottoman and Egyptian governments used to pay huge fees to the Bedouins in return for their services.[56] In reality, the actual aim was to prevent these tribes from looting the hajj caravans. Where payments were delayed or they were dissatisfied, Bedouins compensated for their loss by plundering the caravans. Therefore, they thought they might lose their privileges when the empire became dominant there, and they were right, given that the Young Turks had actually introduced the reforms to spread the central authority to the provinces. Their fees were cut down. The *surra* allowances would now be paid only if they could protect the railway.[57] The payments formerly made to Bedouins to protect the Egyptian hajjis were also reduced.[58]

The Bedouins feared that government centralization policies would disrupt their autonomy and independence in the region. The first signs did emerge when the railway approached their territory. For instance, the Bedouins and railway administrators disagreed about the issue of water supply.[59]

By 1902, the Bedouins had already begun to show signs of rebellion by cutting the telegraph wires of Hejaz and knocking down their poles.[60] As the Hejaz railway approached Medina, similar assaults escalated. In 1901, German consul in Damascus wrote that there was no resistance against the railway construction around Damascus but reactions arose in the south, and that the government would take more measures in that particular area.[61] As a matter of fact, towards the south, Ottoman garrisons on duty along the Hejaz railway had a higher number of soldiers and weapons. The total number of soldiers protecting the railway exceeded 5,000. Station buildings that were protected by the army had almost turned into fortresses. High stone walls were built with only a few loopholes left.[62] Station buildings were more frequent, and wooden crossties that the Bedouins had vandalized were replaced by steel ones. However, even the delivery of steel crossties to the construction site was a challenge. In June 1908, when 22,000 tons of steel crossties were about to be shipped to the port of Rabigh, the manager of the company Düsseldorf Stahlwerk-Verband stated his reservations about possible Bedouin attacks.[63] The tension evinced itself in this decision taken by the Ottoman cabinet on October 27, 1907:

It has been decided that all the trains will be supplied with an armoured wagon at the back, with two machine guns inside, in order to protect them from any possible assault by the Arabs in the desert; eight defence quarters will be built to include two army troops between Medina and Mecca, containing two battle cannons of 7.5 centimetres and two machine guns; the soldiers in these quarters will be armed with rifles, soldiers that inhabit the quarters and block pools will be provided with necessary arms and food supply for four weeks; cisterns will be built in areas where there is no well ... and the total number of soldiers protecting the railway between Medina and Damascus will be increased from 300 to 500.[64]

Despite all the measures, Bedouin assaults, especially by those residing in the south of al-Ula near Hejaz, the Bani Harb and Bani Ali tribes, could not be averted. When the construction between Tabuk and Medina was almost completed in the summer of 1908, the assaults took the form of threatening attacks. A more serious assault came when some Bedouin sheiks were arrested by Construction Minister Kazım Pasha, who had come to Medina in the beginning of 1908 with a group of 1,300 people, consisting of soldiers, engineers and railway workers. Kazım Pasha must have overestimated his power in the region that he dared arrest some of the tribal leaders who had denied him support.[65] After this incident, he left the town with two troops with the purpose of negotiating with some tribal leaders; however, brutally attacked by the Bedouins, he had to take shelter in the al-Hasa region three hours away from Medina.[66] In this attack organized at the Ashar Gate, seven soldiers were killed and 13 were injured. Kazım Pasha reported to the Ottoman government that unless necessary precautions were taken, the Bedouin tribes would dare attack not only the railway but also Mecca and Medina.[67]

This attack triggered subsequent Bedouin rebellions in 1908. The level of violence and aggression gradually escalated. When a total of 3,000 Bedouins raided Biyâr-i Nâzif station near Medina, 72 people, including two army officers, lost their lives. Upon this incident, new troops were sent to Damascus, Jerusalem, Smyrna (İzmir), Trebizond and Macedonia.[68] 1,400 soldiers from Dedeagaç and Smyrna and 1,012 soldiers from Trebizond were sent to Medina. The forces in Medina were increased to 11 troops and the number of soldiers in each troop from 400 to 600.[69]

Similarly, the number of soldiers recruited to the Hejaz railway was increased to 15,000, and the soldiers working on the construction site were also equipped with rifles. Construction work was carried out in groups of 20 to 50, all armed. The Bedouins were strictly banned from the construction zone beyond firing distance.[70] Two wagons in the form of armoured cars were ordered from Europe. The decision to purchase two armoured vehicles was suspended due to the impracticality of their use.[71] In the meantime, the Ottoman authorities decided to start peace negotiations with the sharifs and sheiks and listen to their complaints.[72] Marshal Kazım Pasha noted that on January 2, 1908, he negotiated with some of the influential tribal sheiks, explained to them that the main purpose of the line was to render the hajj more comfortable for Muslims, and assured them that all benefits and gifts granted by the Ottoman Empire would be maintained. The sheiks gave Ottoman authorities an oath of allegiance and sustainable peace. The Pasha also demanded a donation of 500 liras from the government to be delivered to some of the tribes.[73]

In the meantime, some local administrators argued that regular troops would not be sufficient to thwart possible Bedouin attacks. For instance, Commander of the Ottoman forces for Medina Bahri Pasha stated that, "the soldiers who are deployed to contain and, if necessary, fight against Arab desert tribes should be fit for extreme heat and other conditions in the region, chosen from the 'sophisticated soldiers of Medina'".[74] Another Ottoman local notable, Rashid bin Nasır, added that if he had been duly empowered, he could have re-established government authority in the region. He suggested that a total of 1,000 soldiers able to ride camels be selected from among the young men of the Arabian Peninsula and equipped with state-of-the-art firearms. He claimed to have full knowledge of the customs and traditions of the tribes so, if authorized, he would be able to secure peace, make the tribes obey the great authority of the Ottoman state, and thus help the state just as he solemnly promised.[75] A report released by the Amirate of Mecca proposed to mobilize a 450-strong armed force commanded by government-appointed officers. The soldiers were supposed to courageously combat the tribes so that the Jeddah line could be protected properly.[76]

In conclusion, it may easily be inferred that the struggle against the Bedouin tribes failed. The tribes fought using cunning guerrilla tactics and succeeded in undermining the Ottoman forces. For instance, in July 1908, a midnight raid by the Bedouins resulted in the massacre of 300 Ottoman soldiers. In another similar attack, two officers and 33 privates were killed; one officer and 64 privates were injured.[77] Despite all the precautions, many Ottoman soldiers were plundered, stabbed and killed during the attacks and skirmishes.[78]

The fact that a group of Bedouins temporarily invaded the Kaaba in Medina proved that the attacks were not limited to rural areas. The invaders went so far as threatening to keep the holy gates shut until they were given what they deserved, and to prevent Muslims from performing the prayer both on Friday and during their religious festival. Consequently, all necessary precautions were taken in order to prevent such a disaster. Yet, however strong the threats and severe the punishments, the Commander of Medina Bahri Pasha, in a coded telegram, advised the authorities that the only way out was to send more troops and urge the tribes to leave the area, and that he could not be held accountable unless the army took urgent action.[79]

In spite of everything, on August 22, 1908, the first train arrived at Medina. Before moving onto the further events of 1908, let us evaluate the Bedouin attacks from another viewpoint. In 1912 a British agent, A.J.B. Wavell, with the pseudonym Ali bin Muhammed, wrote his memoirs after he disguised himself as a Muslim and travelled across Beirut, Damascus, Medina, Yanbu, Jeddah, Mecca and Yemen. The third chapter of the book, with the title "Hejaz railway", covered the author's train journey from Damascus to Medina right after the line started operating. The Bedouin resistance was depicted with real stories and anecdotes. Wavell, having ventured to dress like a Muslim, for he would not have been admitted to the south of al-Ula otherwise, reaffirms the aforementioned observations regarding the Bedouin resistance:[80]

All the stations south of Medain Salih are fortified with trenches and barbed wire, and the whole scene reminds one of South Africa at the time of the war ... We were told that it was by no means unlikely that we should be attacked ... We therefore looked to our weapons on re-starting ... the dull thudding of distant artillery fire told us that we were approaching our destination [Medina]. The stations were now protected by considerable earthworks and had garrisons of a company or more. As we drew nearer the rattle of musketry fire became audible ...

The part of Arabia being theoretically a province of Turkey ... However Turkey has little real authority in the Hejaz. The Bedou remain, what they always have been, independent tribes, each community having its own country, rulers, laws, and customs. ... The pilgrims consider them savages and have good reason to hate and fear them ...

For many years past the Turks have found it less trouble to pay a certain sum of money to the sheiks of the Bedou tribes through whose country the pilgrim caravans have to pass, in return for immunity from attack, rather than to send large escorts with them. Though it may well be considered undignified for a civilised government to submit to such extortions in their own country, there is really no help for it. To occupy and police Arabia in such a manner as would make it a safe country for travellers, would be at present about as practicable an undertaking an invasion of the moon.

With the completion of the Hejaz railway the Turkish government made a precipitate and, in the circumstances, an ill-advised attempt to stop further payment of tribute for safe conduct to the tribes en route. The news spread through Arabia and alarmed the more important tribes. ... If they were not allowed to plunder and not paid to refrain from doing so they would evidently be in a bad way.

... The Benee Ali on their side proclaimed a sort of holy war against the Turks, and invited all Arabs to assist them ... The assistance they asked for was soon forthcoming ... For once in a way the tribes seemed in perfect agreement. ... At the time of our arrival the Turkish troops in Medina may have mustered

10,000, with twenty guns; the Arabs upwards of 20,000, and were daily increasing.

Wavell's observations are parallel to the information gathered from other sources, which, in the end, may be interpreted as British intelligence agencies' mastery in appointing agents across regions to gather information. In Syria, Mesopotamia and the entire Arabia, specialized agents, like Shakespeare, M. Sykes, Gertrude Bell, Leachman, Bramby, and Parker would travel to the remotest corners and collect information.[81]

At that stage, Bedouin attacks were not directly associated with Britain, yet it was well known that the British sought to provoke the tribes in line with their interests. For instance, Abdülhamid's secretary Tahsin Pasha had reported that the British were provoking the Bedouins.[82] Furthermore, Germany's consul to Beirut wrote that the use of dynamite in demolishing telegraph poles implied British backing behind Bedouin attacks. According to the consul, it was clear how the British, the governor of Mecca and the sheriff mobilized such support.[83]

Parliamentary debates during this period demonstrate the growing hostility among the British against the Ottomans as the rails approached southern Arabia. For example, at the beginning of 1911, Lord Lamington, one of the conservative members of the House of Lords, submitted a parliamentary probe on Ottoman railway constructions in Arabia and their upcoming projects. He cautioned against any potential railway project in the hinterland of Aden, which was still under British hegemony and had great strategic importance. In return, he was assured by Parliament that Britain's policy was primarily based on the preservation of the status quo.[84]

Analyzing the Ottoman government's suggestion to a British shipping company and why it was rejected may help further understand Britain's approach towards the Hejaz railway. The Hejaz Railway Administration had suggested that the Khedivial Mail Line that transported Egyptian hajjis change their conventional hajj route. This would bring certain advantages for the Railway Administration, hajjis and the agency. For the agency, the rise in the share of shipping within its transportation activities would ensure higher profitability. For the hajjis, they would not have to travel between Mecca and Medina by caravan, thereby avoiding highly probable Bedouin attacks. Despite its potential benefits, British authorities asked the Khedivial Mail Line to reject the proposal. British rule in Egypt did not want the Meccan amirs, whom they gave full support, to lose the profit they had been earning on the journey between Mecca and Medina. Moreover, in order to detract from the reliability of Ottoman rule, they believed it was necessary that Egyptian hajjis go through this exhausting 15-day journey.[85]

In brief, all domestic and foreign references were full of evidence that Bedouin attacks and sabotage kept mounting in 1909. The Young Turks cut down on their annuities, which exacerbated their rage and eagerness to fight. That was the primary reason why Hejaz governor and commander Soleman Pasha insisted that "the customary annuities be paid to Arabs regularly". Soleman Pasha stated that "Abd al-Rahman, a former hajj warden, proposed to send hajjis to their destinations by sea, which infuriated the Arabs even more".[86] The commander of Medina stated that those Arabs who rebelled against Ottoman rule were manipulated by "false convictions that those attacks would help them gain the upper hand once they were heard in İstanbul".[87] Furthermore, misleading news about a prohibition of camel transport and new regulation of women's clothing contributed to further escalation.[88]

The Bedouin tribes concentrated their attacks in the south of al-Ula, and acted in unity to resolve all internal disputes.[89] In 1910, workers in the Zarqa district quit work due to insecurity.[90] The Bedouins kept on looting hajj caravans. A hajj group had to return to Medina on their way to Yanbu.[91] Writing in his memoirs of Maan, the orientalist Musil also reported interesting details. When the railway reached Maan, the government suspended *surra* payments, which had formerly been paid to the Huwaytat tribe. As a consequence, 40 of the tribal leaders decided to meet the district governor of Maan. Invading the government mansion, the Bedouins threatened to kill all those who entered the mansion until their money was paid. The district governor saved innocent people by paying a certain amount and asking them to wait until the rest was paid after he had negotiated with the governor of Damascus. In the end, the leaders left the district, once more warning that unless the payment was made by July of that year, they would be back even stronger.[92]

Despite a handful of tribes who obeyed Ottoman authority,[93] the Bedouins were largely victorious. The Ottoman government had failed to establish law and order along the railway, let alone to continue to build it beyond Medina. What is more, the military operations and precautions cost a fortune. In the end, the Ottoman government decided to shake hands with the Bedouin tribes. During the negotiations chaired by the Amir of Mecca Sharif Husayn, tribal leaders voiced their complaints about the financial losses they would incur once the railway was completed. In order to build consensus, the Ottoman government agreed to keep paying their regular *surra* in return for a vow of non-aggression. The construction work carried out to extend the Hejaz railway towards Mecca and Jeddah would also be suspended. In order to pay the tribes, the government decided to charge an additional 100 kurush per passenger in first class and 50 kurush per passenger in third, travelling between Madain Salih and Medina.[94] Starting from this mutual accord until the rebellion of the sharif in 1916, the Bedouin attacks and conflicts in the Hejaz district almost came to an end. The final stage of the construction, however, was never completed.

3- *Resistance of the Amirs of Mecca*

Another barrier to the construction of the Hejaz railway was the attitude of the amirs of Mecca: they opposed any reform that would eventually reinforce the central authority. Although they gave a pro-Ottoman impression, the amirs underhandedly provoked Bedouin tribes to continue aggression. They tried to hamper all central government investments aimed at controlling the region and facilitating information flow.

The power of the amirs dated back to previous centuries. In fact, all amirs of Mecca under the hegemony of the Abbasid and Mamluk sultanates and finally the Ottomans lacked the economic and political power to act autonomously and independently. However, a given ruler in Baghdad, Cairo or İstanbul had to ensure the cooperation and consent of the amirs on certain matters if he wanted to establish control in the Hejaz. Any sharif in conflict with the government could seek shelter with their Bedouin allies in the desert and torture the hajjis on the way, going unnoticed for a long time. Therefore, even during Ayyubid rule, it was customary to motivate the amir of Mecca to show allegiance to the Mamluk sultans through payments and gifts. The rulers of the Abbasid caliphate, Mamluk sultanate or Ottoman Empire could not challenge this tradition. They lacked the power and will to relinquish their control over the holy cities and the region. The fact that Ottoman sultans derived their legitimacy from their political achievements, not from their heritage, verifies this. The generosity shown to hajjis and the residents of the holy land that was equal to or even greater than that of the Mamluk sultans represented a clear source of legitimacy.[95]

One of the most important income sources for the amirs was half of the Jeddah customs profit. In addition, Ottoman administrators sent a total of 25,000 kurush to the amirs annually, under the title *atiye-i hümayun*.[96] Their financial power derived from their collaboration with Bedouin tribes as well. Primarily making a living by camel transport, Bedouin tribes shared half of their profit with the amirs of Mecca. The amirs also charged three liras to each hajji arriving at the port of Jeddah in return for providing safety during their visit.[97] "Safety" was tantamount to non-aggression by the tribes, which clearly illustrated the collaboration between the amirs and Bedouins. In conclusion, the amirs in the region were among the staunchest advocates of the status quo, so that they could maintain their good old power.

Having made a detailed study of the Bedouins, Oppenheim explains in brief why Ottoman and Mamluk governors endured and tolerated the tribes. Simply put, the answer is financial. "An ongoing military protection to the region cost a lot . . . so the system had never been modified, until 1925 when İbn Saud purged it."[98] Similarly, Abdülhamid rejected Ahmet Muhtar Pasha's offer to abolish the amirates, stating that, "if amirs became dysfunctional, holding sway over the Arab public only with appointed governors would be impossible, unless an army corps were deployed there". Abdülhamid asked Muhtar Pasha to be more realistic.[99] In short, Abdülhamid's remarks echoed Oppenheim's argument.

The amirs of Mecca as well as the Bedouin perceived the modernization attempts in general and the arrival of the telegraph and the railway to Medina in particular as the harbingers of change of a system that had been in place for centuries. In order to document the attitudes of the amirs toward the developments that would eventually alter the balance of power in the Hejaz, we could go further back in history and consider the practices of Amir Abd al-Muttalib. He communicated with the British consul and some sheiks in Jeddah in opposition to the Ottoman state. When Osman Nuri Pasha, who was the Hejaz military commander and governor, became aware of his actions, the Ottoman government removed Amir Abd al-Muttalib from his post and appointed Aun al Rafiq Pasha in 1882.[100] Yet the replacement of Abd al-Muttalib by Aun al Rafiq Pasha did not alter the situation at the Hejaz. The new amir first came up with many obstacles to the extension of the Hejaz telegraph line to Mecca. Interestingly, the reason behind the amir's attempts at such blockade appear in a report of the French consul in Jeddah: ". . . hence the reason for the difficulties he created through the Bedouins working at the telegraph line connecting İstanbul to Mecca thus become evident. He knows very well that the day direct connection is established with the center, he will no longer be the effendi in his own house . . . ."[101]

We had mentioned in chapter two that Amir Aun al Rafiq had not given his support to the proposal of the Hejaz military commander Osman Nuri Pasha in 1892 regarding the construction of the Jeddah−Mecca railway. In addition, Amir Aun al Rafiq also tried hard to have the Hejaz governors that did not accommodate him removed from office and was often successful in such attempts. For instance, Cemal, Safvet and Osman Nuri Pashas were removed from the governorship of Hejaz by sultan Abdülhamit in accordance with the wishes of the amir.[102] Among them, Osman Nuri had presented a petition to Abdülhamit suggesting the limitation of the influence and authority of the amirs of Mecca.[103]

As for Ahmet Ratip Pasha who was later appointed to the Jeddah governorship, since he preferred to participate in the self-interested relationships of the amirs, he was able to sustain his post until the 1908 Young Turk rule. Ahmet Ratip Pasha also turned a blind eye to the practices against the Ottoman state of Sharif Ali who had succeeded Aun al Rafiq as the amir of Mecca and as a consequence accumulated significant profits. For instance, the governor Ratip Pasha gained five liras from every camel rented out to the pilgrimage caravans and, given that approximately 100,000 camels were rented annually, this would amount to 500,000 liras.[104] The most concrete example of how the governor sided with the amirs during the conflicts of interest that emerged between the Ottoman state and the amirs of Mecca is his attitude in 1908 during the negotiations held between the sides regarding the Hejaz railway.

The Ottoman government realized that they had to find a way to remain on good terms with the amir of Mecca, and so sent Construction Minister of the Hejaz railway Kazım Pasha and *Yaver-i Ekrem* (Aide of the sultan) Arif Pasha to negotiate with Sharif Ali and Ahmet Ratip Pasha. Arif Pasha explained to the governor that the sultan put great emphasis on the Hejaz railway project and declared that he was being rewarded with the title of Aide to the sultan.[105] Since the amir of Hejaz and the Hejaz governor were in full agreement regarding their relations of self-interest, they had openly avoided supporting the Hejaz railway construction by coming up with various excuses. When Kazım Pasha asked the governor and amir of Mecca their opinions about the railway, the amir asked to defer the construction that year on the grounds that he was sick. Infuriated, Kazım Pasha said, "We cannot get things done with such idleness. For six years, we have been trying hard to finish the line, even working in miserable conditions. I have been sleeping on stones, in dust... Where are those that call themselves loyal? Please get serious!". Hearing those words, the governor and amir were uneasy and held numerous meetings to block the construction.[106] When the correspondence between the Komisyon-ı Ali, Kazım Pasha and the amirate are examined, it is obvious that the commission was aware of the controversies among the parties.[107] The order they sent was succinct: "any misunderstanding or disagreement has to be settled immediately so that the construction may be launched".[108]

Governor Ratip Pasha and Sharif Ali tried to hide the proclamation of Constitutional Monarchy from Hejaz residents. The amir of Mecca ordered that those speaking in favour of the Ottoman Constitution of 1876 be whipped, and with the support of Governor Ratip Pasha, he mobilized the tribes against Constitutional Monarchy. On the other hand, many army officers welcomed the new rule of the Young Turks, formed branches of the Committee of Union and Progress and organized demonstrations in favor of the new regime.[109] In the meantime, when the new government dismissed the governor from his post, a large crowd attacked the governor's mansion. Ratip Pasha was arrested and his assets were seized.[110] Osman Pasha, the commander of Ottoman forces in Medina, was dismissed from his post due to alleged opposition to the new constitutional regime. Meanwhile the local people had already begun to react to the new policies. For instance, an attempt by the Committee of Union and Progress of Mecca to collect municipal tax provoked riots.[111] The new government appointed Construction Minister of the Hejaz railway Kazım Pasha as governor of the Hejaz and Sharif Husayn as the amir of Mecca. Consequently, power relations between the local authorities of the Hejaz and the central government were impaired, paving the way for new actors, new circumstances and new relationships.

The amir, representing the local authority, and the governor, representing the central government, naturally had a tense relationship. When problems were exacerbated, the regime generally withdrew their representative and tried to prevent further escalation. Local people usually supported the amirs.[112] In his report, Ahmet Muhtar Pasha, who offered to abolish the amirate, pointed out that the problem was a systemic one, independent from individuals: "as long as the system remains two-tiered, problems are inevitable, no matter who is appointed as amir or assigned as governor".[113]

In fact, the Ottoman rulers were not desperate. The most important advantage they had was that the sharifs, all descending from the prophet's family, were divided among rival tribes. If the sultans found themselves at loggerheads with a particular amir, they could simply depose him and replace him with another one from another tribe. After 1840, the rivalry between two Hashemite families, Awn and Zayd, contributed to the political turmoil in the Hejaz even further. Many Ottoman governors were dismissed and many sharifs were sent to İstanbul in exile to lead a life of indulgence. When replacing Sharif Ali, the new government had two options: Ali Haydar, one of the leading members of the Zayds, and Sharif Husayn ibn-Ali of the Awns. Since Husayn was not on good terms with the amir of Mecca (his uncle), he had been leading an indulgent life in exile in İstanbul since 1894.[114]

Some researchers believe Ali Haydar was supported by the Committee of Union and Progress, and Husayn was backed by Abdülhamid.[115] In fact the circumstances must have forced them to act fast, so those in the first party could not consider the issue in so much detail. The time when Husayn reached Mecca to take over the office coincided with the time when the first hajjis arrived at Medina by the Hejaz railway. In this period, when Bedouin rebellions occurred, the Ottoman rule had to resort to Husayn's mediatory skills, which added to his power. He could play a helpful role in the eyes of the rule and be nice to the Bedouins, at the same time by rewarding them with regular payments. What is more, he managed to intercept the railway before entering his region, for fear that it could threaten his power.

In order to ensure implementation of their reforms, Ittihadist politicians believed that the amir should not enjoy absolute power. Local members of the committee led by Abdullah Quasem welcomed Husayn with the following words: "We have come to welcome the constitutional amir, who, it is hoped, will disregard the old administrative procedure and the despotism of Sherif Aun al Rafiq, a leader who understands the modern way and the constitutional changes necessary for progress and safety." The amir replied, "The sultan who founded the constitution which you have mentioned is proud that he and his predecessors are the servants of the Holy Houses of the Hejaz; and a servant is not a master. This country abides by the constitution of God, the law of God and the teachings of His prophet."[116] He gave the first signs of the policy he would implement, and that the struggle would be a harsh one.

In 1909, when Fuat Pasha became governor, the tension between those two poles was at its highest. Bedouins also interfered as expected. When Fuat Pasha wanted to end slavery, the tribes threatened to obstruct the way between Mecca and Jeddah. Fuat Pasha had to hold back. Governor Hâzım Beg was appointed to Beirut by the prime minister, due to his disagreements with Husayn.[117] The elections of 1911 also proved how powerful the amir was. His sons Abdullah and Faisal were elected MPs from Mecca and Jeddah.

The government conflicted with the amir on the separation of Medina from the Hejaz amirate. When the amir asked for information on the subject, they said, "Since communication between Medina and the capital by telegraph and railway is now direct, the governorship of Medina may now be considered independent, and directly under the Ministry of Interior and not the Province of the Hejaz."[118] These words made clear that the telegraph line and the railway were vital in establishing governmental control in the region, which the amir must have felt uneasy about. The killing of three Indian hajjis in 1912 and all the other disturbances were interpreted as the amir's own doing in order to take Medina under his control. He tried to create the impression that all the planned reforms during this period would only become possible with his will and consent.[119] The policies of Ittihadists to make reforms and those of the amir to preserve the status quo became the deepest fault line in the Hejaz railway construction process. Once the new government came to power, they desired to take up the project. First, they planned to build the Mecca–Jeddah branch line and calculated the cost as 180,000 liras.[120] After this, payments made by the hajjis, especially from India, would fund the Mecca–Medina section of the Hejaz railway. Equipment for the Jeddah line was imported from Belgium and Spain and stored in Haifa.[121]

Governor Kamil Bey arrived at Hejaz in June 1910 and relaunched the project. In March 1911, 13 engineers in Jeddah began to measure the field. The prime minister ordered the amir of Mecca to take the required measures.[122]

However the opposition of the amir and tribes and the Tripoli and Balkan wars brought the construction to a halt. Due to an imminent Italian bombardment, shipping railway equipment via Jeddah and Rabigh bore risks.[123] The Austrian consul at Jeddah deemed it impossible to build the related lines unless the power relations in the area were shifted.[124]

In 1913, the construction discussions were revived. Prime Minister Mahmut Şevket Pasha wrote in his diary on February 14, 1913, that in a meeting with Minister of Internal Affairs Adil Beg, they decided to garner support from the Muslims of India for the project.[125] In the meantime Abdulkareem Abu Ahmad Khan Ghaznevi from the Indian general assembly had held a meeting with Husayn on the railway, where Husayn for the first time openly declared his opposition. He could agree only if the financial losses of the Bedouins were compensated and he was given full authority on the Mecca line. Ghaznevi informed the Ottoman state about the meeting and urged it to come to an agreement with Husayn, who had great influence over the tribes. He added that the project could easily be funded by all Muslims, primarily Indians, who were eager to see the completion of the railway.[126]

After the Balkan war, Ittihadists maintained their policies of centralization. For instance, they opened party branches in the Arab provinces, published newspapers in Arabic and tried to cooperate with Arabian intellectuals such as Shakib Arslan for regional integration. Shakib Arslan, from the amir's family in Beirut, had always defended integration and unity both during the reign of Abdülhamid and in the Ittihat era, and was in favour of modernization of Islam. He won the elections in 1913 and became MP from Hauran and, once the war broke out, was appointed editor of the newspaper *al-Sharq* published by Cemal Pasha in Syria, having close relations with renowned Muslim philosophers of the time like Muhammad Abduh and Jamal ad-Din al-Afghani.

In 1913, the Ittihadists sent him to Medina to make pro-Ottoman propaganda to Arabs and help eliminate opposition. Arslan was supposed to pay a visit on behalf of *Cemiyyet-i Hayriyye-i Islamiyye,* an association newly established in İstanbul aiming at, according to him, forging good relations between Turks and Arabs. Sultan Abdülaziz's son Yusuf İzzeddin was the chair and Said Halim Pasha, the general secretary. Among the members were Talat and Enver Pasha, together with other prominent Ittihadists. Its composition indicates the association's role in governmental policies. When Ittihadists failed to suppress Arab demands for autonomy, they founded such an organization and sent Arslan to the region with similar aims in mind. Shakib Arslan reported what he did in Medina directly to Talat Pasha, which proved that this association was closely linked with the Committee of Union and Progress.[127]

In Shakib Arslan's letters, the individuals and organizations that maintained separatist movements were reported, together with their reasons for doing so, and the factors contributing to their "actions and betrayal" were analysed in detail. For instance, the French navy's arrival at Beirut encouraged newspapers like *el Islah* and *el Kabs* to defend even more strongly their views in favor of obedience to French rule. Arslan suggested abolishing such newspapers immediately. Furthermore, French influence in the region was to be eliminated and replaced with Ottoman influence by restricting concessions to the French and by accelerating Ottoman-sponsored projects such as the Hejaz railway. The hajjis he interviewed also supported his suggestions and made complaints about the present situation. He knew that the circumstances did not permit them to complete the Hejaz railway as planned, and therefore made another suggestion to satisfy the hajjis in the short run. A line between al-Ula station and al-Vech district could be built and the journey from al-Vech to Jeddah be extended by four or five ferries, which would ensure "the comfort and safety of Muslim pilgrims and their belongings". Gaining the hajjis' sympathy would boost his reputation. He also urged paying three or four months' worth of salaries to the inhabitants of Medina, which had not been paid for eight months.[128]

In his second letter, Arslan repeated his suggestion: "If enmity, a shame for the world of Islam, is eliminated, the number of hajjis will rise to 500,000 from 120,000." His words prove that he made an effort to deter tribal sheiks from opposing the Hejaz line. According to him, many Arab sheiks promised to contribute to the extension of the Mecca–Medina line if the remaining part of the land was officially granted to them.[129] In his third letter, he claimed that the khedive of Egypt intended to take over the caliphate—and thus Arabia—with the help of the British. He proposed some mechanisms to prevent this. In addition to the sheiks who had been persuaded, some others sent messages to the mediators declaring they would agree if the profit they made out of camels was compensated.

Arslan also delightedly declared his readiness to found a branch of *Cemiyyet-i Hayriyye-i İslamiyye,* with an invitation sent to 60 of the most prominent men. He also noted the existence of two other Islamic organizations, *Himayetü'l-Ka'be* (consisting of 60,000 men from India) and *Hareket-i İslam* (from Java). For him, the Ottoman state could benefit from these organizations to rally financial and spiritual support.[130] In Java and Sumatra there lived 35,000 people who were loyal to Ottoman rule but suffered from the pressure of Holland, and *Hareket-i İslam* would lead them to dwell in Islamic cooperation. The Ottoman state, therefore, had to support these organizations in terms of providing *hodjas* (religious instructors) and other means of education.[131]

As a result of Ittihadists' efforts to permeate both the provinces and their legal codes with central authority,[132] Husayn's reaction against Ottoman rule mounted. The amir's reaction seems natural given the fact that Vehip Beg, newly-appointed Hejaz governor and commander of Ottoman Forces, attempted to fortify the Ottoman garrison in the region and render the Hejaz a province ruled by law. Vehip Beg was a radical Ittihadist, believing that the only way to suppress disobedient practices in the region was to display a decisive and authoritarian attitude. In February 1914, with this aim in mind, he arrived at Hejaz with seven troops of foot soldiers and one of artillerymen. Soon, the tension between Vehip and Husayn became obvious. In 1914, during the hajj, Vehip Beg's army deployed on the mount of Arafat outmanned that of the sharif. He aimed at founding a charity to help legal officers not to ask for financial support from local notables.[133] The new governor also revoked the arms granted to the sharif's private Bedouin wardens.

There was no doubt that Vehip Beg was determined to execute provincial laws and extend the railway towards Mecca. In the end, on February 23, 1914, Husayn sent a coded telegram to the Prime Ministry, asking for a replacement of the governor.[134] In his telegram, he noted that the governor seriously hindered his efforts to secure law and order. In the meantime, Hejaz inhabitants and Bedouins began to react more. Bedouins' obstruction of the road between Mecca and the coastal parts of Medina led to starvation in the cities. When the government asked them to agree, the amir visited the governor, finding a large crowd in front of the governor's mansion. In his talk, he clearly opposed the execution of the laws: "You can see for yourself the desire of the Hejaz people to maintain their former rights which Sultan Selim the First gave them when he was appointed caliph." He stated that they would by no means collaborate unless the decision to apply district laws to this local area was withdrawn. In the meantime the crowd outside protested against the elimination of privileges and the extension of the railway: "No railway from Medina to Mecca!", "Long live the Amir!", and "Down with the Law of Provinces."[135]

In return, Vehip Beg proposed radical solutions, such as the dismissal of the amir and the deployment of two big divisions. First, the dismissal of the amir was accepted, but later on Prime Minister Sait Halim Pasha hindered Talât Pasha (Minister of Internal Affairs) on the grounds that it might cause great chaos, and that the loan to be granted from France might be at risk.[136] Let us now quote the words of the Commander of Medina, Fahrettin Pasha, commenting on these events:[137]

> The tribes began to gather and make heavy demands from the new rule, following [Husayn's] propaganda that he had launched by sending letters and soldiers to Arab tribes (on the grounds that as a result of the new Mecca–Medina line, the new governor would lay off camel riders, recruit more soldiers to the

army, and starve people to death). Soldiers traveling from
Jeddah to Mecca fell to an ambush, six were martyred, five
injured. . . . Similarly, on March 9, 1914, Husayn revealed his
real ambitions with a coded telegram sent to the Prime
Ministry in the form of an ultimatum: As seen from the
telegrams sent by the public (to Internal Affairs), the
opposition is now crystal clear. Disastrous events on the roads
and in cities (soldiers killed and injured) have risen to a level we
cannot cope with. Disagreements are predicted to lead to worse
consequences. . . . The sherif promised to provide peace in the
region if the proposal were accepted.

Finally, Amir Husayn won his political struggle against Vehip
Beg, forcing Ottoman rule to officially announce that the railway
would not be extended any further. In the telegram immediately sent
by the Prime Minister, the Ottoman government assured the amir
that their privileges would be restored and the extension of the
railway would be halted. Once this telegram was read out loud in the
mosque, law and order were restored.[138]

Talât Pasha had also attempted to reach an agreement with the
amir, who interviewed Abdullah (the amir's son and deputy of Mecca)
and asked him about recent events in the Hejaz. Abdullah clearly said
that the problem arose from the government's desire to abolish the
Hejaz's autonomous status:

If you want the Sherif to serve the government in the Hejaz on
the basis of true Muslim brotherhood and the Sherif is the
guardian of the heart and its arteries. The government's best
interests would be more truly served by leaving the old
constitution in force rather than applying the new measures.

Hearing this, Talât Pasha asked him why Husayn was against the
construction of the railway. He replied that his father's original
intention was not stopping the construction; however, once the
Ittihadists came to power, the objectives of the railway as explained
by Abdülhamid were modified:

The completion of the Hejaz Railway would take the livelihood away from those who live on camel transportand by acting as guides to the pilgrims. The concern of the Sherif today must be your concern as well, namely the building up of a Muslim policy, with its centre in the Hejaz and the Sherif as its custodian.[139]

Although Talat Pasha was not content with this reply, he at least conveyed the right message to Husayn via Abdullah. If they managed to cope with the opposition to the railway, Talat Pasha assured them that they would receive one-third of the railway profits and all soldiers working in the construction would be under their command. Husayn would also receive 250,000 pounds to deliver to the Bedouin tribes, and his sons' inheritance of the amirate would be guaranteed. He also stated that the amir would be dismissed from his post if he disagreed. Both Abdullah and the sherif rejected the offers, claiming that this was nothing but humiliation and bribery. The sharif had already said that he had had no other concern but the safety and peace of the holy lands.[140] When World War I broke out, the issue was suspended due to the new circumstances.

Vehip Beg had displayed the same radical attitude towards Husayn, at the expense of contradicting the conciliatory approach of the Ottoman rule. He sent report after report complaining about the amir, his arbitrary practices and judgements, his court being paid without any labour, miserable conditions in jails and Mithat Pasha's provocations against the government. İstanbul asked for tolerance and conciliation, which, Vehip Beg knew, would not work at all. He ended up asking for Husayn's arrest on the grounds that he was working against the government and his two sons' being taken into custody in the capital. İstanbul asked for tolerance and conciliation, again. This time, he insisted that the situation was unbearable, and asked the government to dismiss either himself or Husayn. He even claimed that Husayn would not hesitate to cooperate with the enemy in case of a raid over the Red Sea.[141] An oracle that actually would come true in the years that followed, Vehip Beg tabled another interesting proposal. He proposed that the government either arrest

Husayn or withdraw from Arabia for good. For him, the Hejaz front line had secondary importance in the war, without any direct effect.[142] Therefore, "there is no need to pursue control over a land that is impossible to defend under current circumstances".[143] In fact, there was earnest truth in his words, which will be supported by writings from Lawrence and Ali Fuat Erden in the following chapter.

C) FRENCH OBSTRUCTION OF THE AFULA–JERUSALEM LINE

On September 17, 1908, Chief Engineer Meissner Pasha suggested that a branch line from Afula—at the 36th kiloeter of the Haifa–Dera line—to Jerusalem be built, claiming that it would bring great financial gain.[144] Extending 113 km, the Afula–Nablus–Jerusalem line would pass through a highly populated and relatively fertile area, which also attracted a high number of tourists. Beirut's Provincial General Assembly also referred to the line in a report: "Every year more than 100,000 tourists visiting this area to go to Palestine would bring within three years a profit that would eventually make up for the construction losses, and even exceed it." The delivery of agricultural equipment would also be easier, and new agricultural devices would lead to a more fertile Palestine. It was suggested that the line be built as quickly as possible, since "once the land gets richer, farmers will profit from this, obtaining the power necessary to struggle with the Jews."[145] In fact, in 1907, the Komisyon-ı Ali also made reference to the economic and military use of the line.[146]

Another factor that would increase the number of travellers is that it would provide cheap and easy transport from Jerusalem to the area around Tabariya lake,[147] where nearly 15,000 Christian pilgrims travelled every year. With a branch line, Muslim pilgrims would visit the tomb of Salahaddin al Eyyubi in Damascus and proceed to the Mosque of Omar in Jerusalem.[148] Many goods such as olives, oil and soap exported from Nablus (with a population of 30,000 people) were distributed via the Jaffa–Jerusalem railway: now this operation could be shifted to the Haifa–Dera railway. Since the Haifa port had more room for development compared to Jaffa, the profit to be gained from the new line would exceed the profit that the French made out of the present line.[149]

There were other factors. The present Damascus–Medina railway would only be useful and popular for hajjis and only on the south of Amman, and transport of voyagers and goods would fall to a minimum outside the hajj season. This would lead to redundancy and idleness. What is more, equipment such as tracks and traverses were kept waiting in Haifa for the extension of the railway. In fact, no construction would be launched from Medina to Mecca in the near future.[150] If the Afula–Jerusalem railway were to be built, all this equipment left to decay in stores could finally be used, and the railway workers be employed. Otherwise the technical staff would also have to be laid off since the construction of the Jeddah line was also not planned then.[151] In addition to all these factors, this line ought to have been supported by the military, as well.[152]

According to the Hejaz Railway General Management's report dated July 1, 1912, the military was about to start arranging the land for the railway, for the Afula–Cenin section.[153] When Deutsche Levante Linie got the news that the construction was being launched, it asked the Ministry of Foreign Affairs to investigate the potential opportunities that this line might bring to German industry. Since construction equipment would be shipped via the port of Haifa and that the shipment would be carried out by them, they were indispensably concerned with the matter.[154]

By 1913, only a 17 km part of the line from Afula to Cenin had been built.[155] Later on, the line could extend only to Nablus. The reason why the Afula–Jerusalem line was not completed has been mentioned before. The French did not wish to tolerate any competitors in the region. They had the economic and political power to hinder them. As previously mentioned, they had also prevented Hejaz railway managers from being a competitor on the Haifa–Dera railway, reducing prices against the Beirut–Damascus–Muzeirib railway. We may easily recall that the reason why the Ottoman government had yielded to their power was the financial difficulties they had had and the loan contract they had with the French. In 1913, another sacrifice the Ottomans had to make was that they stopped the construction of the related line, in order not to compete with the Jaffa–Jerusalem line.[156] The French were not satisfied with this, and went further to demand the benefit to be gained from the Afula–Jerusalem line, which they asked to be linked to their railway.[157] In conclusion, the obstruction of this line, which was actually highly profitable and easy to build, led to frustration and resentment on the part of Hejaz railway managers and officers against the French—and the Ottomans, for succumbing to their demands.[158]

Cemal Pasha summarizes this situation as follows: "In order to gain profit out of this business, the French demanded such concessions that they seemed to aim at our lives." The principal condition was that the Ottomans should not build any railway line within Syria and Palestine,[159] which shows how important it was for the French to strengthen their economic and political influence there.

All in all, the Ottoman State was deprived of any chance to build railway lines, even on their own land. The British in Aqaba, the French in Syria and Palestine, the amirs and Bedouins in the rest of the land starting from Medina left them with no choice but to stop the extension of the Hejaz railway in all directions.

D) OTHER UNREALIZED LINES AND SOME ADDITIONS

Let us now move on to miscellaneous ideas to extend the Hejaz railway to Yemen and Baghdad. In fact, long before the Hejaz railway project, in 1889, Abdülhamid had mentioned the idea of building an additional railway in Yemen, between Hudaida and Sanaa. In August 1889, he had conveyed a message to Deutsche Bank through the German Ambassador. The Bank's General Manager Siemens (building the Baghdad railway) evaluated the Yemen railway project, focusing on the fact that kilometers might not be guaranteed and the profit gained from coffee in Yemen would be inadequate. When Wilhelm II was also interested, an engineer was sent to Yemen; however, when problems concerning the Baghdad railway could not be solved, the Yemen Project had to be deferred.[160]

The Hudaida–Sanaa project was once more raised after 1904.[161] What they had in mind was to link this line with the Hejaz line in the long run, which would connect Yemen to the center by train. On June 5, Mahmut Şevket Pasha said, "We shall begin to build the Hudaida–Hacile line with Ottoman capital and link it to the Hejaz railway. When the railway İstanbul–Sanaa is completed, we shall rule Yemen like other provinces and eliminate all local authorities in this way."[162] The German military attaché in İstanbul also complained about the weak political position of the Ottomans in Yemen, and mentioned the importance of the Hudaida–Sanaa line, and the extension of the Hejaz line across Yemen.[163] German writers supported the idea with the assumption that the region would be linked to the Baghdad railway. If the railway were linked to Yemen, a new trade route would emerge between the Indian Ocean and Europe. The religious purposes of the line would be overshadowed by commercial ones.[164]

In 1905, a French engineer named Jaborowski was commissioned to prepare a Hudaida–Sanaa railway proposal. During that year, a group of French engineers came to the area to investigate. As a result of financial difficulties and pressure of the French, the Yemen Railway was saved for a French company. Pre-construction was completed in 1911, and the railway was planned to spread over 328 km in the next year.[165] The infrastructure of the first 50 km was also completed. However, a series of reasons such as the line's being damaged under Italian attacks during the Tripoli War, the Bedouins' obstruction led by Imam Yahya (who was backed by the British), and the strategic importance of the railway for the British, led to the line's construction being halted.

The last factor mentioned above played a great role as evident in Austrian diplomatic documents of the time. For instance, Robert Deutsch stated that the Ottomans desired to build this railway so that the rebellions led by Imam Yahya, provoked by the British, could be contained: "It would make it faster to deliver soldiers from the coast inland. Up until then, due to Bedouin attacks and geographical difficulties, 10 per cent of soldiers lost their lives." According to the reporter, unless the railway was built, "rebellions with both internal and external origins cannot be stopped, the required reforms cannot be put into practice, trade and transport cannot be developed."[166] Apart from Deutsch, the British who settled in Aden and believed that Sanaa was under their authority were also aware of this fact. This time, they tried to take Austria on their side. This was planned as another method of obstruction. Deutsch reported that British officers told him that, "Newly flourishing French authority in Yemen should be terminated, and the French Company be replaced by an Austrian equivalent." Imperialism clearly pursued different goals in different parts of the world.[167] The interesting thing was that the reporter did not totally ignore this suggestion. In the rest of the report, he considered what type of strategy should be followed in case the railway was left to their authority.[168] As usual, the war brought the construction plans to a halt.

From time to time Ottoman authorities had been planning to extend the Hejaz railway over Najd to Baghdad. For example, Mahmut Şevket Pasha wrote on his diary: "I have worked on the Baghdad–Mecca railway. We set our minds to build it, but the Germans and the British also intended to. It was much more proper for us to do the construction."[169] Despite all the efforts, not even the Baghdad and Hejaz lines could be completed. In this case, it might seem unreasonable to design a railway connecting the Hejaz to Iraq. Nevertheless, the Germans, in line with their Arab policies, endorsed such projects that would eventually reinforce Ottoman hegemony.[170]

We should also note that some minor additions were made on the Hejaz railway before World War I. In 1911, the starting point of the railway was shifted from Kadem (4 km away from Damascus) to the city center. In 1913, another 17 km line from Haifa was built, which linked the Haifa–Dera branch line to the city of Acre.[171] Shuttling between the two coastal cities, three trains left every day from each station and the journey took 45 minutes. The connection of Acre to the Hejaz railway was made possible by the help of a prominent notable of the city, a member of the Committee of Union and Progress and an MP candidate from that party, Sheik Esad.[172] According to him, unless something was done urgently, the Muslim city of Acre, bearing the highest potential in the region, would not be able to compete with the Christian/European city of Haifa. If there was any social vitality at all to Acre, it was due to the presence there of the office of *mutasarrıflık* (district governorship). The transfer of the governor's office from Acre to Haifa was hindered by Sheik Esad, as well.[173] With the connection of the branch line from Acre, the lively social structure of Haifa would not diminish anyway, thanks to many banks and ship agents.[174] The Acre line was also vital in military and strategic terms.

Another branch line built on the Hejaz railway was opened in 1913, among Dera–Basra–Old Damascus. A district with a high potential for wheat production was now able to join the market, and Druze rebellions could now be quelled.[175] 35 km long, costing 30,000 liras, the line was contracted to Ottoman builders and the construction was carried out by Ottoman workers under the supervision of Ottoman engineers.[176] This led to the French authorities' speculating again about how unfairly their DHP Company had been treated[177] on the grounds that the Dera–Old Damascus railway would attract the products of Hauran to the Hejaz railway.

Plate 1   An Ottoman Committee visiting Dera Station, 1905.

Plate 2   Railway Construction in Damascus district, 1901.

Plate 3 Ottoman Operating Officers traveling by private train on the Hejaz railway, 1905.

Plate 4 The Construction of Maan railway line, 1903.

Plate 5 The opening ceremony of Haifa Station, 1905.

Plate 6 Tabuk Station, 1909.

Plate 7   El Ula istasyonu, 1909: Al Ula Station, 1909.

Plate 8   The opening ceremony of Al Ula Station, 1907.

Plate 9   Al Akhdar Station, 1909.

Plate 10   Hajis traveling by Hejaz train, 1909.

Plate 11   Dera Station, 1903.

Plate 12   An Ottoman Committee departing from Maan to Damascus, 1905.

Plate 13   Train carrying construction materials for Hejaz railway, 1905.

CHAPTER 6

WERE THE EXPECTATIONS FULFILLED?

A) ADMINISTRATIVE DIMENSION

1- *Hejaz Railway Staff*

The Hejaz railway was the only railway built and operated by the Ottomans. As we know, the initial goal was to run the project without foreign capital and to manufacture all railway equipment at local factories. The Ottomans could only partially succeed in this by raising funds for the project from fellow Muslims. As discussed in chapter 4, efforts to promote local production had failed and eventually all of the necessary materials had to be imported.

Recruiting only local Muslim staff at construction sites was another failed attempt. However, at least some top-level staff were chosen from among Muslim Ottoman subjects. We have already explained that new engineers graduating from the *Hendesehane-i Mülkiye* were employed on the Hejaz railway and that some of them were sent to Germany to improve their skills. Nevertheless it was foreigners, especially Germans, who undertook the technical management of the project. These foreigners were senior experts who had worked in various railway companies within the Ottoman Empire.[1]

Similarly, in jobs such as carpentry that require a medium level of qualification, Europeans were recruited. Local Arabs were employed as unskilled labour as conductors, firemen, and station chiefs. This led to discordance between Muslims who occupied lower-rank positions and non-Muslims who were their superiors.[2] As time passed, Muslim Ottoman citizens began to work at all levels. The parts of the Hejaz railway where non-Muslims were forbidden to enter (between al-Ula and Medina) were successfully completed by Muslim engineers and other Muslim staff.

The transfer of power to the Young Turks created major problems in Hejaz railway management and construction. Since the new government had serious misgivings about Abdülhamid's approach, they made radical replacements within the railway staff. The foreign staff was dismissed from top-level duties.[3] Meissner Pasha was among those who were laid off.[4] At the end of 1909, when they realized that they could not find replacements for these people, they started hiring foreign staff again. Most prominent managerial and technical positions were again occupied by foreign experts, primarily Germans. For instance, Zaehringer, chief engineer of the Anatolian Railway Company at the time, was appointed as chief Manager of the Hejaz railway. Having worked in the railway industry for many years, he was experienced and qualified enough.[5] In his job interview, his self-confidence and positive attitude had apparently influenced General Manager Cevad Beg.[6]

In conclusion, the principle of employing exclusively Muslim staff was replaced with a more hybrid recruitment strategy. For example, between 1905 and 1912, those who worked as Hejaz railway Chief Manager consisted of three Ottomans, two Germans, and one Frenchman: namely, Irfan Beg, Haji Muhtar Beg, Salih Beg, Gaudin, Zaehringer and Dieckmann.[7] The last of these received a salary of 30,000 francs. Given that 1,000 francs was enough to live on in Haifa with a family, this salary was more than satisfactory.[8]

Reports by German diplomats in the Ottoman Empire kept referring to the continuous tensions between Ottoman and European top-level managers. For instance, the consul in Haifa, Hardegg, once claimed that manager Zaehringer and senior engineer Nötel were about to resign although, in fact, their contracts were valid for much longer. According to the consul, they were highly beneficial for the railway. More interestingly, Hardegg predicted that the new manager might resign soon, on the grounds that the position had too much responsibility with no authority. The manager, who lived in Haifa, was a mere technical advisor. Hardegg presented frequent replacements at senior level (five different managers since 1905) as evidence to this fact. He also wrote about Dieckman, who had worked on the Berlin–Hamburg railway and was briefly employed on the Hejaz railway in 1908.[9]

There were also a lot of problems concerning maintenance and repair operations. In 1905, Hardegg reported that before the construction was complete and the line began to operate, the parts had broken down, because necessary precautions had not been taken. In 1910, General Manager Cevad Beg complained that only 44 out of 71 locomotives were ready to operate, while some others were not ready for a long journey. He was concerned that because the lines passed through the desert, sand would fill the mechanical parts and they were not competent in solving sand-induced problems. The Medina and Madain Salih workshops had not begun to operate in full capacity.[10] Austrian traveller Musil reported that only 13 out of 71 locomotives were able to operate. Only one out of a hundred railway workers in the area had mechanical know-how. Metal elements of brand new wagons were stolen and sold. The railway administration could not guarantee the safety of the travellers' belongings.[11] Along the Hejaz railway, main repair shops were located in Damascus, Haifa, Dera, Maan and Madain Salih.[12]

Between Damascus and Medina, trains ran three times a week. Additional trips were added on religious holidays and special occasions. Trains departed from Damascus at 7.30 a.m. and returned on the same day at 3 p.m.[13] With an average speed of 25–30 km/h, trains would travel from Damascus to Medina in 72 hours, which was later reduced to 55.[14] The same path had taken 40 days by camel. The cost of the journey was also reduced to 200 francs from 1,200.[15]

The architecture of the Hejaz railway was a projection of the dual objectives that the project was supposed to fulfil: military (e.g. fortresses, junction stations and military barracks), and religious (hajj accommodation and mosques).[16] The departure and arrival times were synched with prayer times. Voyagers could perform their prayer in special glass wagons custom-made at the *Tersane-i Amire* (Grand Shipyard). In 1909, even an imam was assigned. First- and second-class wagons for four persons could be divided into two parts to ensure a comfortable journey.[17] Moreover, the third-class wagons had female-only compartments.[18]

2- *Analysis of the Operating Revenues*
The main section of the Hejaz railway, the Damascus–Medina section, was used for the transportation of goods and passengers only up to Amman. The distance between Damascus and Amman was 222 km: the journey from Amman to Medina, which was 1,080 km, was cost-effective only during hajj seasons. The Haifa–Dera line (161 km), however, could compete against the Beirut–Damascus–Muzeirib (DHP) line thanks to the agricultural fertility of Hauran. For instance, 20,148 passengers and 14,836 tons of goods annually were transported via Haifa station. Samakh was another busy station, with 10,108 passengers. The only three stations along the main line of the Hejaz railway where the number of passengers exceeded 10,000 annually were Damascus, Dera and Medina (46,229, 11,674 and 24,506 respectively).[19]

An analysis of annual revenues and expenditures of the Hejaz railway reveals that it was a profitable undertaking. For instance, in 1910, an income of 267,891 liras was recorded while the expenditures totalled 196,725; which equalled a profit of 71,166 lira. However, it was a public investment, no interest had to be paid. Had that not been the case, the profit made in 1910 would have equalled only 1.95 per cent of the capital—3,899,197 liras— and this would have been marked as loss.[20] The interest rate was approximately 4 per cent. In 1909, 1911 and 1912, the profit was 34,069, 82,936 and 98,174 liras respectively.[21]

The Hejaz railway differentiated itself from other railways in that, while other railway companies in the Ottoman Empire usually made a profit out of goods transportation, the Hejaz railway did so through passenger traffic. For instance, in 1912 income from passenger traffic amounted to 167,571 liras while that of goods transportation was

84,896 liras, 55 per cent and 28 per cent of the total profit respectively. 39.9 per cent of the profit was from hajjis, while 15.5 per cent was from other passengers. This trend was maintained in subsequent years. For instance, in 1911, profit from passenger traffic was 175,289, that is, more than half of the total profit.[22]

If the Hejaz railway travellers are grouped into two, hajjis and ordinary passengers, then hajjis constituted only one-sixth of the total travellers. However, they accounted for 70 per cent of the administrative profit. This complication resulted from the fact that the hajjis travelled all the way from Damascus or Haifa to Medina, while the other passengers travelled shorter distances. For instance, in 1913, only 32,650 out of 240,716 travellers were hajjis (13.6 per cent). However, 70.6 per cent of the total profit (130,600 out of 184,929 liras) was made thanks to hajjis.[23]

Ticket prices also varied on different parts of the line. For instance, on the Haifa branch line, first-class tickets per kilometer were almost 0.50 kurush, second-class per kilometer was 0.375 kurush and third-class per kilometer was 0.25 kurush. Different prices were offered along the line between Damascus and Medina. For instance, in the northern part of the line (between Damascus and Dera), third-class tickets were reduced to 0.20 kurush in order to compete with the French DHP Company. On the other hand, military safety measures taken on Dera–Medina line increased the prices; first-class, second-class and third-class ticket prices per kilometer being 0.60 kurush, 0.45 kurush and 0.30 kurush respectively. Passengers to Medina had to pay an additional 20 kurush in return for health care service provided in international quarantine centers in Tabuk.[24]

Goods transportation (per kilometer and per 100 kilograms) cost 0.55 kurush for first, 0.50 kurush for second, 0.45 kurush for third-classes. In some regions, however, prices were reduced in order to compete with DHP and camel transport.[25] For instance, between Haifa and Damascus where there was fierce competition with the DHP Company, 130–159 kurush was reduced to 62–100 per ton. In ports like Yanbu and Rabigh, where goods were exported and carried to Medina by camel, prices were reduced to attract trade to the Haifa line.[26] However, camel transport could not be eliminated altogether despite these efforts.

B) RELIGIOUS/POLITICAL DIMENSION

As already mentioned above, thanks to the Hejaz railway, the journey from Damascus to Medina was reduced to 55 hours from 40 days, while the cost dropped by one-sixth. The line also eliminated the trouble and exhaustion of travelling by camel and being exposed to Bedouin attacks. Quarantine centers and healthcare measures helped control contagious diseases to a certain extent.[27] As a result, 8,777 hajjis used the line over continues long distances to the last stop Medina in 1909 and 7,079 in 1910, and the number reached 13,102 in 1911. In the meantime, if we add to this the number of hajjis who travelled by Hejaz railway over short distances we get the numbers 14,965 in 1909, 25,079 in 1910, 29,102 in 1911, 30,062 in 1912 and 32,650 in 1913. In other words, the number of hajjis using the line between 1909–1913 had more than doubled.[28]

The Hejaz railway was the culmination of the policies that Abdülhamid had formulated in order to prevent the virus of nationalism from infecting the Islamic elements of the empire. Is it plausible to argue that the Hejaz railway had served its declared purpose? In other words, did the railway help increase the number of hajjis and eliminate the troubles and dangers they had previously been exposed to?

The aforementioned figures clearly reveal that the Hejaz railway could not entirely fulfill its religious function. From 1909 to 1912, an average of 9,445 hajjis travelled by rail.[29] Figures cited by Austrian consul Toncic (9,000 hajjis annually) confirm this.[30] Toncic reported that during five years following the opening of the Medina line, 160,000 of those who went on hajj from the north used sea transport, while 45,000 travelled by train. He emphasized how the Hejaz line had failed to meet religious expectations. Given the fact that during the Sacrifice Festival, 250–400,000 hajjis gather in Mecca, the ratio of hajjis travelling by train was relatively low. In 1911, notwithstanding those who reached Mecca through the Desert of Arabia[31]—nearly 150 thousand hajjis the number of those making the hajj via the Hejaz railway was 13,102 people, that is, 13.5 per cent of all hajjis. In the same year, 83,822 pilgrims used sea transport and reached Mecca by caravans from Jeddah and Yanbu. Those who travelled by sea were 58,147 people from eastern countries such as India, Iran, Iraq, and Java; 13,586 people from northern countries such as Syria, Anatolia, Rumelia, Russia, Turkmenistan, and Tatarstan; 12,089 people from Egypt, Morocco and Algeria. In 1909, the number of hajjis that preferred the Hejaz railway to sea transport was 8,777 (10.9 per cent), in 1910 7,079 (7.1 per cent) and in 1912 8,823 (8.3 per cent).[32] Obviously, an overwhelming majority of hajjis (approximately 90 per cent) preferred to reach Mecca by sea.

One reason why they chose to travel by sea was that the railway between Mecca and Medina was not yet completed. That is, those who went to Medina by train had to travel further (450 km) to reach Mecca, with a lot of difficulties. Hajjis naturally did not prefer to travel on camels, under the sun and the danger of Bedouin assaults. Instead (except for those who travelled from Syria) they preferred to reach Jeddah by boat, and then complete the journey to Mecca (74 km) on camels alongside other caravans. In this case, since the part of the Hejaz railway between Medina and Mecca was not completed, the whole line failed to fulfill its function of facilitating the journey for hajjis. The incomplete railway between Jeddah and Mecca created even bigger problems. Statistics indicate that half of all hajjis reached Mecca through Jeddah.[33] If this short line of 74 km had been successfully built, then at least all of the hajjis who preferred the sea voyage would have travelled to Mecca safely and comfortably.

After fulfilling their hajj duty in Mecca, many hajjis went to visit their Prophet's tomb in Medina. Another evidence of this fact is that the number of pilgrims who used the railway on the return journey exceeded those who used the railway on the way there. For instance, in 1909, 8,777 Muslim hajjis used the railway on their way to reach the place, while this number rose to 11,188 on the way back. In 1910, those who returned by the Hejaz railway amounted to 18,000. In the same year, only 7,000 hajjis travelled there by the Hejaz railway.[34] The figures overlap with those reported by the Jeddah consulate.[35] It is obvious that a part of the hajjis reached Mecca by boats, then continued their journey with a visit to the Prophet's Tomb, and joined the caravans in Medina on the way back. Since the challenging parts of the journey were already left behind, they took the train from Medina to Damascus or Haifa.

This also explains why Indian Muslims were interested in the Mecca–Medina or Jeddah–Mecca parts of the railway only. If these two lines had been built, an Indian hajji could have first reached Jeddah by boat, come to Mecca by train after a very short journey, then used the railway for the tomb visit and travelled back by train once more through Jeddah. The fact that these lines had not been built led to a loss of prestige in the eyes of Indians. In fact, expectations were high at first. For instance, a newspaper article in 1904 wrote that when the Hejaz and Baghdad railways were completed and linked to Riyaq, Indian, Afghan and Middle Eastern hajjis would not have to go to Jeddah from Bombay through Aden, having the chance to reach Basra from Karachi in four days by the sea, and move on to Riyaq, Medina and Mecca through first the Baghdad and then the Hejaz railways.[36] Since neither of these lines had been completed, these expectations were never fulfilled. On duty in Mecca on behalf of the sultan, *Yaver-i Ekrem* Arif Pasha, wrote in 1908, "this railway, if not completed between the two Harams [Mecca and Medina], shall turn out to be useless". He kept his hope for the near future: "If the line, hopefully, is extended towards Yemen, then the Arabian peninsula will be connected to İstanbul by land. In 2–3 years, once Aleppo and Konya are also linked, it will be possible to take the train from Haydarpaşa and travel all the way to Yemen."[37]

In 1912, those who went on hajj by Hejaz railway were 8,823, while this number reached 19,000 on the way back; and the total number both to and from was distributed among countries as follows: nearly half (48.6 per cent) of those who used the railway on the way to the holy places were Muslims from Russia (4,288 pilgrims), while 20.9 per cent were Egyptians (1,845 people), 11.6 per cent were Anatolians (1021 people), 7.7 per cent were Indians (675 people), and 3.9 per cent were Syrians (347 people). In 1912, those who used the railway on the way back were 7,418 Turkmen, Tatar, Muslims from Bukhara and Russia (39.1 per cent), 4,532 people who travelled back to Anatolia (24 per cent), 3,000 Iranians (15.8 per cent) and 1923 Syrians (10.1 per cent). Railway passengers on the way back included 596 Iraqis 326 Algerians and Tunisians, 304 Egyptians and 156 Muslims from Rumelia. Few hajjis used the railway line from Bosnia, India, Afghanistan and other Muslim countries.[38] Only 304 out of 1,845 Egyptian hajjis used the railway on the way back, simply because of the tough quarantine processes they had to go through imposed by the British in Egypt.[39]

During the hajj season of 1910, some hajjis, fed up with waiting due to unavailable tickets, preferred to go back through Yanbu, which put the Railway Association into trouble. General manager Cevad Beg complained that there were not enough enclosed wagons on the Hejaz railway, and that they did not get any answer from the Ottoman rulers for their various requests for more wagons. To solve the problem, they used open wagons (called low-edged platforms) to carry hajjis, which later banned by Ministry of Health. Permission was granted after covering open wagons with sun shades, but still, only 400 hajjis could be transported. Cevad Beg claimed that as long as they deferred the supply of enclosed wagons, difficulties would continue.[40] His warning proved right: in 1912, many hajjis could not use the railway during the return for there were no available seats. The Railway Administration was desperate. Every day a train (with a capacity of 350 passengers) left Medina, but could not meet the demand. 4,000 hajjis visiting the prophet's tomb in Medina were left without any seat. They had to return by ship from Yanbu.[41]

Another problem the Hejaz Railway Administration had to cope with was impoverished hajjis' attempts to travel free of charge. Fugitive hajjis, when caught, were thrown from the train. The Ottoman government was heavily criticized because of the cruel way they treated these poor hajjis, which lowered their prestige. A Bosnian wrote an article in the newspaper *Haji al-Asr'ul-Cedid*, saying:

"Inspectors charge the hajjis without any tickets double price, and hajjis who cannot afford tickets are forced to leave the train sometimes in the middle of the desert. A short while ago, a poor hajji thrown from the train by an inspector was tragically torn to pieces under the wheels. ... The poor devoid of money or tickets were either left out in the desert, exposed to all kinds of dangers, regardless of whether they had the food, drink, and capability to survive under such severe conditions; or were cruelly thrown out of trains, in the most inhuman way." The governor of Syria conducted an investigation and declared that railway managers had defended themselves by saying they were simply obeying the rules. He then requested "a legislative amendment to avoid similar problems and bring these shameful and disastrous events to an end".[42]

General Manager Cevad Beg had actually claimed that forcing free riders to leave was a part of railway regulations, although he agreed that the implementation was harsh. Nevertheless, he accepted that the practice had to be abandoned because it could lead to unpleasant outcomes in small stations in the midst of the desert. Cevad Beg feared that showing tolerance to fugitive voyagers would eventually encourage other free riders. Moreover, giving the railway officer too much leeway in the implementation of rules could result in abuse of authority. Looking for an interim solution, the Council of Ministers declared that 3 per cent of hajjis would be transported for free, and began to arrange free journeys for 250 poor hajjis.[43] However, the inspectors were instructed to check all the other passengers, apart from those constituting the 3 per cent.[44]

Another report by Cevad Beg addressed the same problem: "If free travel becomes common practice, both the railway line and the state treasury will incur a remarkable financial loss, which is unacceptable." He also remarked that Hejaz railway prices were the cheapest across the empire.[45] The problem remained despite the three per cent pro bono practice. A report by the commander of Medina proves this fact: "18,000 hajjis were sent to Damascus by train, three per cent of whom amounted to 540 people. Our administration has sent them to Damascus pro bono so far. However, we are expecting more than 40,000 hajjis this year, and 200 more poor hajjis need to be and will be transported to Damascus."[46] In another report, he asked for authority to send poor hajjis back to their hometowns free of charge. He was told to charge them half price. Hajjis had to either sell their belongings or borrow money to return. Commander of Medina Basri Pasha said, "We keep announcing that free travel is now prohibited, but they keep on reproaching and yelling at the municipality ... Poor hajjis are almost 2,000."[47] Apparently, the Ottoman government was in deep trouble. It would be inappropriate to make the journey available only to those who could afford it, for the raison d'être of the line was to extend a helping hand to hajjis. On the other hand, 3 was no doubt that money was needed for the line to operate. A quota of 3 per cent did not solve the problem.

The part of the Hejaz railway to the south of Maan was only used during hajj season, on religious holidays and occasions. For instance, on the Day of Mevlûd, birth anniversary of Prophet Mohammed, Muslims visited his tomb. In 1912 and 1913, 1,219 and 2,229 people reached Medina by the Hejaz railway respectively. In the same way, on Laylat al-Miraj, when Mohammed was believed to have ascended into the heavens, in 1912 and 1913, 1,020 and 3,228 people used the railway respectively.[48] The Hejaz Railway Administration offered discounts on fares to encourage people to travel to the holy lands on special religious days.[49]

In a nutshell, the Hejaz railway did not and could not permanently eliminate the problems and the dangers the hajjis encountered. The main reason was that the Medina–Mecca and Jeddah–Mecca lines were not constructed. In other words, the Hejaz railway did not facilitate the hajj journey, and hajjis kept using sea transport instead.

C) MILITARY/POLITICAL DIMENSION

1- *Pre-War Period*

As we already know, military and strategic factors played a significant role in the conception of the Hejaz railway project. First of all, the line would enable rapid soldier deployment to the region without the need to use the British-controlled Suez Canal. The Ottoman government believed they could reinforce their power in the Hejaz and Yemen thanks to the railway. Besides, the line was on the road to India, which obviously added to its strategic significance. The line also increased prospects of victory against British-ruled Egypt. After the de facto invasion of Egypt, the British had believed that the place was naturally under their rule due to geographical reasons. The Hejaz railway made them feel uneasy and insecure. If the Baghdad railway could be completed, the Persian Gulf would be controlled by Ottoman/German rule, undermining British influence in the Arabian Peninsula.[50] But to what extent were these ambitions really fulfilled?

Obviously, the Hejaz railway indeed speeded up the soldier deployment process. As soon as the railway had reached Maan in 1904, it began to be used for military purposes. For instance, the Bedouin riot that broke out in spring 1904 could easily be quelled by 30,000 soldiers sent to Maan by railway.[51] In 1905, the riot in Yemen provoked by Imam Yahya could be contained thanks to 28 troops that were immediately sent from Syria. It took them 24 hours to travel from Damascus to Maan, a journey that formerly took 12 days. However, after Maan, soldiers had to travel a long and difficult road. It took four days to travel the 110–120 km from Maan to Aqaba near the Red Sea,[52] and another five days to reach Yemen. This instance reveals that if the British had not blocked the Maan–Aqaba line, it would have gained considerable military value. If the line had been built, the journey would have taken 4–5 hours instead of four days.

In 1910, another riot by the Druze was immediately suppressed thanks to rapid soldier transfer. Another evidence of increasing safety in the region were the villagers of Tabuk (a settlement with nearly 400 residents), who had been displaced due to Bedouin attacks but returned home after the railway had been built. Tabuk inhabitants were farmers who had to abandon their land due to assaults by the Beni Atiye tribe.[53] However after the station had been opened, a hospital, school and mosque were built, creating a more vibrant atmosphere.[54]

The Hejaz railway strengthened Ottoman authority around Syria, especially in the villages around Dera, Maan and Amman. It also helped decrease Bedouin attacks and boosted security in the region. Increased Ottoman authority in the region meant obligatory military service, tax obligations and Ottoman monopoly on arms delivery to local people. They had not been accustomed to such obligations and regulations. For instance, in 1910, when the Ottoman state, which had not hitherto mobilized any soldier or tax from the region, attempted to hold a census in Hauran and Karak, to recruit soldiers and collect weapons from the public, people protested harshly. A riot broke out in Karak, 30 km away from the Hejaz line. 3,000 Karak settlers rebelled against the Ottoman government.[55] After this riot, Bedouin tribes in the region attacked Quatrana station, damaged the station building and killed railway officers. Thanks to the troops that were dispatched, the riot was quelled.[56] However, similar attacks continued in the years to come.

Annual statistics concerning the military staff and goods carried via the railway reveal that 1910 was a year of boom. The reason for this is the riots mentioned above. In 1909, 8,480 soldiers were carried by railway, while this number reached 77,661 in the following year. 766 tons of goods carried by trains in 1909 reached 4,225 tons in 1910. Once the riots were terminated in 1911, the number of soldiers decreased to 27,390, and tons of military materiel to 2,183. However, delivery of soldiers spiked once more in 1912 and reached 47,941 soldiers.[57]

In a nutshell, as soon as the line reached Maan, the Hejaz railway began fulfilling the function of delivering soldiers. Although Ottoman authority in Syria was strengthened after the opening of the railway, it was still not strong enough to impose military duty and taxes on the locals. The Ottoman hold on the Hejaz, however, was not as strong. A report written by the Command of Medina provides evidence of the fact that the situation was highly desperate:

... In order to protect desert settlers from possible attacks from the hills, a troop of camel riders is present all the time. The soldiers here are under the threat of attacks, not to mention extreme heat during the day and cold at night. Far from the stations, they cannot communicate. Any news of an imminent attack cannot be directly conveyed, and even if it were, these

platoons would definitely fall into an ambush due to desert
conditions. Therefore, police stations are required to provide
security ... I have been concerned with and trying hard to
launch the construction and complete it in a few months, for
the sake of the safety of hajjis, security of the soldiers and
honour of the state.[58]

2- *War Period*

In order to better grasp the military importance of the Hejaz railway,
it is crucial to examine its function during World War I. Upon the
closure of the Suez Canal, the Ottoman government's sole contact
with Syria, Palestine and the Hejaz was the Hejaz railway. Syria and
Palestine were provided with troops and Medina was supplied with
food thanks to the Damascus–Medina railway. Due to the war, it was
impossible to send food to the Hejaz by sea, so 20,000 sacks of wheat
and flour were sent from Damascus to Medina by train.[59]

During the war, new regulations were adopted to make better use
of both French railways and the Hejaz railway. After the war had
begun, French lines were confiscated and bound to the Ministry of
Defence. An officer of the French railways in the region protested
this, saying that being at war with the French did not and should not
entitle the Ottomans to take over all the railways. He claimed that
although the railway companies had shares in French companies, they
were legally Ottoman companies. According to him, the Ottomans
would definitely need foreign investment after the war as well;
however, such an approach would damage their credibility in the eyes
of European investors.[60]

Due to the war, all railways in the empire were taken under
military control. In 1916, the Hejaz railway was also tied to the
Ministry of Defence.[61] New lines began to be built for military
purposes. Already in fall 1914, a branch line had been built in
Damascus, linking the Beirut–Damascus–Muzeirib railway built by
DHP and the Hejaz railway. So it became possible to depart from
Riyaq and arrive at Medina without any transfer. Moreover the
Haifa–Dera line started running at night.[62] Hence, in 1914 and
1915, the railway was used at night as well, during the military

operations over the Suez Canal. For two weeks, nine trains full of soldiers and equipment travelled to a location near Jerusalem where troops began marching towards the Suez Canal passing by Beer Sheva. When the operation failed, the soldiers withdrew over the same route via the Hejaz railway again.[63]

Throughout World War I, 437 km of new lines were built for military purposes. At the beginning of 1914–15, the line starting from Afula station situated on the Haifa–Dera railway and proceeding towards Jerusalem finally reached Nablus. This new line would originally be linked to the Jerusalem–Jaffa railway, but construction was later cancelled. Orenstein & Koppel Company's proposal to build a railway from Maan to the Sinai Peninsula was also rejected.[64] Instead, the Hejaz–Egypt railway from the Afula–Nablus line towards Egypt would be built. The railway would begin in Mesudiye, go to Beer Sheva through Lid, and finally reach Egypt.[65] "Enver Pasha and Liman von Sanders decided that Meissner should once more be appointed to the region for this duty."[66] At the time Meissner was working on the Aleppo–Islahiye section of the Baghdad railway. Praising him, Cemal Pasha told Enver Pasha that "this appointment would be most appropriate." Cemal Pasha had met Meissner during his tenure in Baghdad, and upon realizing his expertise, he believed he had better work for the Egypt–Hejaz railway, due to its higher strategic importance.[67]

German diplomatic documents indicate that negotiations and pre-construction work had already begun between German and Ottoman authorities before the Ottomans entered the war. For instance, a report dated October 19 stated that the Hejaz–Egypt railway was intended to assist in the invasion of the Sinai Peninsula; it would be built by German engineers led by Meissner; five months were needed to complete it; Meissner was working in Baghdad at the time, but was planning to arrive in southern Palestine in 21 days.[68] Taking this document into account, it can be claimed that both Ottoman and German authorities already knew the Ottoman Empire would go to war.

Under extreme conditions of war, equipment (especially racks and traverses) necessary to build the Egypt–Hejaz railway became hard to supply. That was why Meissner Pasha decided to use the spare traverses of the Baghdad railway then sitting in workshops in Damascus, customizing them for the Hejaz–Egypt line. Since the Baghdad railway was wide and the Hejaz railway was narrow gauge, adaptation was necessary. Once the adaptation proved inadequate, less vital tracks from the Haifa–Acre and Damascus–Muzeirib lines were removed and used for this new line.[69] The 28 km Dera–Bosra–Eskihan railway was also removed for the same purpose.[70] As predicted, railway troops were employed for the construction, with a total number of 450 soldiers. Another group of 1,500 soldier-workers, and an additional 700 soldiers would join later on.[71] The Hejaz–Egypt line was given utmost importance for military reasons. In the report by Dr Jaeckh to Foreign Affairs, the Baghdad–Hejaz railway, from İstanbul all the way to the Sinai Peninsula and Suez Canal, was referred to as vital for Germany. He wrote his report after meeting with a group of German technicians and experts, including Helfferich, the Minister of Finance, and Moltke, German Chief of Military Mission to the Ottoman Army.[72]

If this railway had been completed up to the Suez Canal, it would have contributed to Cemal Pasha's military expedition to the Canal. In addition, the center of the empire, the Hejaz and Egypt would have been connected by railway, and military equipment and materials would have been deployed faster without having to use the Suez Canal. However, the railway could reach only to Beer Sheva, 25 km from Ismailia on the Suez coast. Ceremonies were held at Beer Sheva Station where Meissner delivered the inaugural speech and the Fourth Army Commander Cemal Pasha drove the last spike, opened the line, and rewarded high-performers with medallions. Senior Ottoman officials, press members, the German consul to Jerusalem, Jaffa and Damascus, the Austrian consul to Jerusalem, and representatives from the German colonies attended the ceremony.[73] General Manager of the Hejaz railway Dieckmann reported that, despite all the challenges they had encountered during the construction, the Hejaz-Egypt railway maintained its everlasting quality.[74] Another reason why the line was considered vitally important was that it had provided a direct railway connection between Jerusalem and Damascus. This was easily achieved by increasing the railway width from 1 m to 1.05 m on the 46 km section of the Jaffa–Jerusalem railway, which had belonged to the French before the war.[75]

The Egypt–Hejaz railway started in Beer Sheva, ran through the desert and finally reached Kuseyme in February 1916. However, repeated Ottoman defeats by the British precluded its construction up to the Suez Canal. When the second Canal expedition in 1916 led by Cemal Pasha failed again, Ottoman/German forces had to remove the line. Otherwise, the line would have been seized by the British, who had begun to counterattack. Still, the only parts that were successfully removed were wagons and locomotives; the tracks and traverses were confiscated by the British. This time it was the British who began to build a new line from Kantara to Palestine in order to counterattack.[76]

Due to their great strategic importance, the Ottoman government tried very hard to keep the Hejaz railway and its extensions operating during the war. Since the Allies had occupied the entire Syrian coast, it was impossible to send coal to the region. In the summer of 1918, weekly reports sent from Aleppo and Damascus read: "No coal left."[77] Coal extracted by German engineers from the Lebanese mountains had low calorific value,[78] and their level of sulphur could easily harm the engines. Wood began to be used instead. However, if only wood had been used, total wood demand would have reached 150,000 tons per year.[79] In the summer of 1916, all trees and bushes around railway lines were cut. New lines were built towards forests just to supply wood.[80] Cemal Pasha told Fahri Pasha, "If necessary, I am entitled to, need to and obliged to even destroy private property in Medina to supply the wood necessary to run the trains."[81] Thorny bushes from the deserts and olive trees in Palestine were also collected to supply fuel.[82] Connecting the Taurus and Amanos mountains on the Baghdad railway made it possible to deliver 100 wagons of coal from Germany.[83]

The gravity of the fuel shortage created another conflict. In December 1916, 60 Austrian–Hungarian employees came from Budapest to work at the railway. This bothered German employees and technicians.[84] In a very short time, the two groups fell into controversy. In the meantime, German engineer Zelle had been trying to supply Lebanese coal but failed. At this point Hungarian engineers took hold of the situation and showed how the coal could be used after being processed in a specific way. Consequently, Germans felt they were losing authority, so they began to seek excuses to blame the Hungarian engineers. According to Austrian Embassy reports, Meissner Pasha wrote against the Hungarians to the Komisyon-ı Ali in İstanbul. German staff aimed at undermining Hungarian influence by finding ways to dismiss high-ranking officials and taking the others under control.[85] In the meantime, two Hungarian engineers, Hajdu and Kiss, came to İstanbul to express their opinions, although Chief Manager Dieckmann did not permit them to do so. In return, Dieckmann sent a report to Cemal Pasha, explaining the incompetence of Hungarians and fired them. However, Hajdu and Kiss persuaded Hakkı Pasha to talk Cemal Pasha out of it.[86] To make matters worse, Dieckmann had asked these two engineers to sign a document declaring that they were not under Gemran pressure, in order to reappoint them. Ranzi, the Austrian consul in Damascus, wrote on April 14 that because the engineers rejected this, they would not be reappointed. He also complained about how the railway administration had divided Austrians into two factions, and that Vegh Milojkovich did not extend the support due to his own citizens. Similarly, Military Attaché Pomiankowski complained that these disagreements were detrimental to the Austrian mission. [87] Austrian diplomats, in their reports sent to the Eempire, asked to what degree they should interfere.[88] Eventually the discord was resolved. Ismail Hakkı Pasha announced "his gratitude for working with Hungarian railway experts now and possibly in the future".[89] According to Ranzi, these incidents detracted from German manager Dieckmann's reputation.[90]

In order for the trains to keep running, they needed oil. They tried to generate oil out of dead animals, animal bones, olive and castor oil.[91] In a telegram sent in 1916, Cemal Pasha announced that the soldiers in Medina would starve to death if oil were not supplied. There was a shortage of spare parts, as well, leading the repair shop in

Damascus to stop operating. Working locomotives began to decrease. In 1917, due to Bedouin attacks, many locomotives were left inactive. The existing wagons were thus overused, causing a decrease in the speed of trains and increase in the length of journeys. Deliveries from Damascus to Medina, normally available in three days, took weeks towards the end of the war.[92] Yet despite all the difficulties, the Hejaz Railway Administration succeeded in keeping the line active until the end of the war.

3- *Arab Rebellion and the Hejaz Railway*
Besides the technical problems mentioned above, the Arab rebellion that broke out in 1916 not only increased the need for railways in an unprecedented manner, but made their protection more challenging. Sharif Husayn maintained his relations with the British since 1912 through his son Abdullah. First of all, Kitchener, the British High Commissar of Egypt, visited Abdullah when he was hosted by the khedive. During this meeting, both parties tried to weigh each other's intentions and made steps towards further collaboration. The first meeting was sincere and positive, and Abdullah was impressed by Kitchener's strong character. According to Abdullah, Kitchener asked questions about the Ottomans' determination to control the holy lands and about relations between Sharif Husayn and the governor, while Abdullah remained cautious in his answers. When they met once more in Egypt two years later, Abdullah openly talked about the tension between the Ottoman government and the amirate, and asked whether British authorities would help them, since the Ottomans were likely to depose the amir. Kitchener's answer was brief and vague. He only said he would bring this matter to the attention of his government; however, his warm attitude and body language convinced Abdullah that he would help.[93] During the same period, Kitchener had the chance to meet the sharif, who had visited Egypt on his way back from İstanbul, and learned that he was highly disappointed with his meetings with the Ottoman government and their decision to extend the railway to Mecca.[94] In his report to British Commissar Edward Grey, he said that Governor Vehip Beg and the amir were on bad terms because of the Hejaz railway, and that

the public supported the amir. Therefore, they still stood behind their demand to suspend the construction of the Jeddah–Mecca line and demote Vehib Beg, who was the governor of Hejaz.[95] In another report, Kitchener mentioned the centralization policies of the Ittihadits as the source of the tensions.[96]

In the meantime, the leaders of the Arab revolts in Damascus consulted Abdullah and asked him to be the leader of the movement led by his father. For the first time in 1914, Abdullah explicitly shared with his father the idea of an independent Arab state. While doing this, he undoubtedly relied on the blessings of Kitchener, the British and Arab leaders in Syria and Iraq. With their endorsement, he surmised, they had the chance to found a powerful Arab state over a vast territory.[97]

When the Ottoman state entered the war, shaking hands with the Arabs was a more attractive option for Britain. On October 31, 1914, the day on which the Ottomans joined the war, Kitchener sent a further message to Abdullah promising that if the Arab nation assisted them in the war, Britain would ensure that Arabia would be spared from plundering. The appalling part of this message was the promise given the sharif that he might become caliph. As we will see more clearly in the next part, while British diplomats hesitated to take any responsibility about the borders of an Arab State, they offered Sharif Husayn a position so highly critical for Muslims, which, in the end, was totally controversial. During the ensuing negotiations, this very offer would outweigh any promise for the borders. According to Kedourie, this conflict derived from the conception of caliphate in Kitchener's mind, since he may have presumed it to be of symbolic value as such a position might have been in the Christian world.[98] Always careful not to offend the Muslims in India and leaders in Arabia, and never taking sides on matters concerning Islam, Edward Grey did not interfere this time, which was equally appalling. According to Friedmann, Grey must have confused the concepts of caliphate and sharifate.[99]

1915 was a year when some important incidents took place that encouraged Husayn to side with the British against the Ottoman state. First of all, the British attacked the coast of the Red Sea, rendering it even more difficult to maintain Arab cities. Besides, Husayn took hold of some documents that contained the Ittihadists' plans to force him to resign, which were delayed with the outbreak of war.[100] Upon this, the sharif sent his son Faisal to İstanbul to talk to the grand vizier. On his way to İstanbul, Faisal visited Damascus and met with clandestine Arab communities to see if they would give support in case of a possible protest. In a meeting at the end of March, the Arab leaders promised to provide support, since three divisions in the city consisted of Arab soldiers. However, when Faisal visited Damascus on his way back from İstanbul on May 23, 1915, their attitude had changed. In a period of two months, governor of Syria Cemal Pasha had contained these communities, arrested their leaders and executed many of them.[101] The three Arab military divisions in the city were dissolved and replaced by Turkish ones, and military officers were exiled to distant divisions like Gallipoli. Cleveland declared that Cemal Pasha's harsh approach led to the "arousal of hatred against Turks".[102]

The few leaders that survived Cemal's executions declared that they would not instigate any more rebellions, and asked Husayn in the Hejaz to turn to the British for help. The leaders also prepared a document drawing the borders of independent Arab territory and gave it to Faisal.[103] With this document, called the Damascus Protocol, it was claimed that the entire territory to the east of Egypt except for Aden would belong to the prospective Arab state. In return, the new state would sign a defence accord with the British, and grant them great economic concessions. Evidently, Husayn had kept the negotiation margin too high, knowing that the British would bring it down anyway. When the protocol was conveyed to the British, they were astonished at Husayn's boldness. Nevertheless, British High Commissar in Egypt McMahon did not want to offend Husayn and offered to postpone the border agreement until the end of the war. After much correspondence, the British left the issue unresolved and showed no willingness to settle. Later on, British political authorities felt they had managed to maintain a successful policy by "postponing the matter and not making any great promise". For example, Gilbert Clayton stated that "Luckily we have been very careful indeed to commit ourselves to nothing whatsoever."[104] Only Lawrence regretted breathing that deceitful atmosphere "with shame and with pain".[105] His confession was a sentimental one that resulted from his emotional ties with Arabs and did not reflect his country's realpolitik or interests. In fact, the deception was mutual. After the agreement, the British realized that Husayn could mobilize neither a sizeable army nor any other support within his secret community. The claim that they could deploy tens of thousands of Arabs for their cause was nothing but fantasy.[106]

More interestingly, the leaders of the Committee of Union and Progress had not seen any of it coming until the rebellions broke out. In fact, the sharif sent his son Faisal to İstanbul in 1915 to persuade the sultan of his father's loyalty. However, some officials in the region had already realized that Husayn and his men were preparing a rebellion and warned the government. For instance, Defender of Medina Fahreddin Pasha had warned Cemal Pasha about the preparations.[107] One of the prominent men of *Teşkilat-ı Mahsusa* (the underground wing of the Progress and Union Party), Eşref Kuşçubaşı, also reported:

"Arab agitation in Syria and all the events leading to the Hejaz rebellion had already been simmering. One and a half years before the rebellion, we had realised the fact that Husayn would clearly act against us once he had the chance. ... Thus we had advised the matter to Cemal Pasha, Enver Pasha, Esat Şukayr, and the Center of the Committee of Union and Progress."

Similar arguments were made by the commander of Ottoman forces in Medina, Basri Pasha.[108] Despite all these warnings, Cemal Pasha did not want to believe Husayn would start a uprising and kept on corresponding with him in a friendly way.[109] The sharif and his sons managed to deceive the government with great skill in two important issues: they pretended to endorse the caliph's call for jihad against the Allies on November 11; and they pretended they would send voluntary soldiers to the Sinai expedition.[110]

Sharif Husayn's letter dated April 18, 1916 reveals that decisions were finally made. As a consequence, on June 9, Ali and Faisal intercepted the trains in Medina and attacked Ottoman locations. Husayn made a declaration on June 27, 1916, claiming that he was right in starting the uprising, that the Ittihadists had restricted the sultan's governmental rights and that secular law reforms and executions in Syria could be damned.[111] One by one, he invaded Mecca, Jeddah and Taif. However, it is impossible to interpret these uprisings as an accomplishment on his side. Although Ottoman military forces were weak in the Hejaz, Husayn's small army turned out to be even weaker. The first assaults were defeated by the Ottoman army and the artillery, and British ships and planes had to help Husayn's troops over Jeddah. The British realized they had to extend massive military and financial support to prevent defeat.[112]

Since Sharif Husayn's alliance with the British made it necessary to open another front, Ottoman army forces had to be divided into two.[113] Now that the connection between forces in Syria and Yemen was also cut off, keeping the Hejaz railway active became more vital than ever. Medina, the last Ottoman territory in the Hejaz, could only survive thanks to the railway. Since the Suez Canal was also closed, the only way to deliver soldiers and arms to Medina, Sinai and Palestine was via railway. The Defender of Medina Fahrettin Pasha's first precaution to protect the trains from Bedouins was to prepare two trains strengthened with sand bags to replace armoured wagons, together with a repair wagon to fix broken trains. Foot soldiers on those wagons were armed with machine guns and cannons. Later on, soldiers along the Hejaz railway numbered 25,000 and machine guns and cannons were deployed at strategic spots.[114]

Lawrence managed to unite the most significant tribes between Mecca and Medina. According to him, Ottoman settlements along the Hejaz railway were surrounded by deserts and unfavourable conditions, while all Arabs living in the desert were in separate homes between vast sand hills, hence able to fight without returning home for weeks. They were like hunters hunting their prey, defeating the enemy in a sudden shock of terror. However, Lawrence also noted that Arabs would not be able to win a regular war fought by regular troops. That is why Lawrence avoided hand-to-hand combat against the Ottomans due to the characteristics of the Bedouins, whom he referred to as "desert men".[115] For him, attacking and demolishing the Hejaz railway was the smartest solution. Lawrence's words set out his war strategy clearly:[116]

> We must not take Medina. The Turks have been harmless there ... We wanted him to stay at Medina ... Our ideal was to keep his railway just working, but only just, with the maximum of loss and discomfort. The factor of food would confine him to railways ... If he tended to evacuate too soon, ... then we should have to restore his confidence by reducing our enterprises against him. His stupidity would be our ally, for he would like to hold as much of his old provinces as possible. This pride in his imperial heritage would keep him in his present absurd position.

Ali Fuat Erden, commander of the Fourth Army conducting the Canal excursion, as it were justified Lawrence's words:[117]

> We were the ones who made Lawrence famous. Our strategies and policies, our perseverance not to surrender Medina and the Hejaz railway turned Lawrence into a "fairy tale hero," a title undeserved. Assume that we send a diver into the bottom of the Pacific Ocean ... Assume that the diver is supplied with air, water and food through a pipe. The entire pipe is attacked by sharks from the surface of the water to the bottom. Nevertheless, all kinds of damages and demolitions are repaired and the diver is left there for years, having to survive against all kinds of challenges. This diver is Fahreddin Pasha. The garrison of Medina. His windpipe is the Hejaz railway. Sharks are the insurgents with dynamite.

These comments by Lawrence and Ali Fuat Erden bring to mind Hejaz Governor Vehip Beg's earlier suggestion to totally withdraw from Arabia. The comments also reveal that maintaining the Hejaz railway turned out to be terribly expensive for the Ottoman government. In Erden's words, "The Hejaz railway was also the railway of the Palestinian front. The power of the Palestine front was correlated with the weakness of the Hejaz front. Each delivery to Hejaz was undermining the Palestinian front ...."[118] Perhaps, as already suggested by Hejaz Governor Vehip Beg, the Hejaz front should never have been opened.

Bedouins naturally attacked the soft spots with massive force, avoiding regular combat and implementing the best guerilla tactics.[119] The British underpinned these attacks with arms, equipment, food and clothes. "The British granted 20 million gold coins, which equalled two billion Turkish liras." Sharif Husayn and his sons delivered this gold to tribal leaders abundantly. Half a century later, they asked a Bedouin leader if he remembered Lawrence or not; the answer was: "He was the one who brought us gold."[120]

Bedouins frequently attacked the railway, damaging tracks and bridges, exploding locomotives and wagons. With the arrival of British experts like Lawrence, the tribes began to wield explosives and dynamite. In a mere four months in 1917, 17 locomotives were damaged. Within the same year, 230 Bedouins led by Lawrence attacked Abulnaim station, damaging a 40 meter track and killing four wardens. Bridges on the north of Maan were also attacked by Bedouins. Wood stocks were targeted during this attack, sparing wood enough for two weeks.[121] After those attacks, all the tracks, traverses and telegraph wires that were damaged or removed had to be fixed, so repair wagons were made available. They tried hard to fix the damage and keep the railway operating.[122] Keeping the trains running between Medina and Maan became a matter of pride and courage, and civil workers on the Hejaz railway were also armed.[123] The Commander of Ottoman Forces in Medina Fahreddin Pasha gave valuable gifts to train drivers and train officers for operating the train safe and sound.[124]

Medina in particular could be defended thanks to the Hejaz railway. If the railway had stopped running, neither the city (surrounded by desert) could have been protected nor the soldiers been able to resist the Arabs' attacks, since their connection with the armies in Syria would have been cut. However, due to a decrease in the number of trains, the food, drink and other material to fulfill daily needs was not supplied in the required amounts. As a result, at the beginning of 1917, civilians began to be sent to Syria, and after a while, the only people left in the city was the Ottoman garrison. Wooden buildings were knocked down to supply fuel for trains.[125] When the war ended, Medina was handed over to the Allies on January 10, 1919.

Overall, the Hejaz railway played a vital role throughout World War I, in the delivery of soldiers, in the military expedition to Suez led by Cemal Pasha, and in resistance against Arab attacks. However, because it was partially incomplete, it could not fully serve its military purpose. First of all, soldiers could not be dispatched directly from İstanbul. In fact, in 1907 the section between Hama and Aleppo linking the Hejaz railway to the Baghdad railway had been completed, but the gaps on the Baghdad railway between the Taurus and Amanos mountains made transport difficult. This problem was only to be solved at the end of the war. Until October 1918, not a single train reached Aleppo. Narrowing of the tracks rendered Riyaq the third connection point after the mountains.[126] Apart from these, the incomplete section of the railway between Maan and Aqaba hindered military action, especially in Cemal Pasha's Suez expedition. Meissner Pasha had raised the importance of this section as soon as the war began; however, it was General Liman von Sanders who hindered the construction of the line. Liman von Sanders stood against the project because of the challenges associated with it.[127]

D) ECONOMIC DIMENSION

It was discussed previously that Ottoman rulers had no economic aspirations when they first decided to build the Hejaz railway. As a matter of fact, the railway brought no direct economic profit or vitality to the region. The line became widely used only to the south of Amman and only during the hajj season. The region's population was quite low. An average of 1,570 people per kilometer used the

Anatolian railways; this figure dropped to 230 for the Hejaz railway.[128] The population of the Hejaz was only 300,000, most of whom, as we already know, were Bedouins. Nearly a quarter of the population lived in Mecca and Medina. Jeddah, with a population of 20,000 people, may also be counted. Both before and after the Hejaz railway, foreign trade, although limited, was conducted via the port of Jeddah. In other words, with her coast near the Red Sea, Jeddah was the Hejaz's most important trade center.

Import goods, namely grain, rice, sugar and cloth, were delivered to caravans from the port of Jeddah just as before. Export goods, primarily leather, fur, palm and resin continued to be brought to Jeddah from internal regions on camels. The main reason for this, at first, was the lack of a line linking the Hejaz railway and Jeddah. However, it should be noted that the Hejaz railway could not eliminate the caravan trade as much as expected. For instance, Medina could sustain itself thanks to caravans coming from Damascus. The reason why people did not abandon caravan trade despite the Hejaz railway was its low cost and the huge distances between train stations.[129]

Every year, during the hajj, economic life in Mecca, Medina and Jeddah gradually became more vibrant. Most hajjis reached Jeddah by sea, visited Mecca for hajj and then some continued their journey by visiting Mohammed's tomb. Those holy visits took a few months every year and the related cities made a lot of money by supplying food and accommodation.[130] For instance, nearly all the Meccans rented a part of their homes during hajj season.[131] Another source of income was working as guides, under various titles such as *shayhk*, *dalil* or *mutavviph*. "Hajjis who did not speak the language of the region had difficulty in finding their way without a 'dalil'.[132] The Austrian consul in Jeddah summarised this situation saying, "agriculture and industry are unknown to the people here, since they make a living only by providing service to hajjis, both in direct and indirect ways".[133] Consequently, the railway was seen as beneficial for local inhabitants, simply because a rise in the number of hajjis meant higher profits. Although Bedouin resistance persisted, some tribes slowly started selling goods like meat, milk or cheese to railway workers and passengers, thus making money.[134] Some observers argued that this would finally lead to their integration into the new centralized system the Ottoman government had aimed at. Similarly, Austrian Ambassador Pallavicini argued that once Bedouins' prejudice against the railway ended, new mechanisms could be mobilized to help them integrate into the new Ottoman system.[135]

Although limited to a few days a year, the Hejaz railway had revived commercial life in Damascus. Returning to their homes from the holy cities via the Hejaz railway, pilgrims stayed in Damascus for a few days and did a lot of shopping. Hajjis not only bought new goods from shops in Damascus, but also sold or bartered the ones they had brought from Arabia. For centuries, Damascus had earned a great deal of money from hajj trade. Even the city was structured in tune with this hajj trade was more common on the way to Mecca. In Al–Sınaiyya district, markets were established to cater to hajjis.[136] The Hejaz railway naturally brought dynamism to the hajj trade in Damascus.

Connecting distant towns to more vibrant markets, Ottoman railways contributed to agricultural development.[137] This was not the case with the Hejaz railway. Once the Haifa–Dera–Damascus line was opened, the agricultural products of Hauran, particularly wheat, were easily transported to the cities and the Mediterranean coast. This led to an increase in the value of the product, but not in the number of new agricultural fields. In fact, arable lands were limited to the north of Amman.

It was widely believed that the area covering the south of Syria and east of Jordan could have become the breadbasket of the empire provided that problems of security, transport and irrigation were resolved. The population of the region needed to increase as well. For instance, in order to accommodate farmers around railroads, new settlements had to be built. Many orientalist scholars claimed that the area was very fertile in ancient times, so improving irrigation facilities would result in a boost in agriculture.[138] However, irrigation projects were never put into practice due to the shortage of labour and capital. Potentially fertile land continued being used as grazing land by Bedouins. Auler mentioned the high number of lemon and date trees in al-Ula district (so abundant that all the Bedouins put together would not be able to consume them), which were not used properly for the same reasons.[139]

An interesting 38-page report written to the Prime Ministry by Max von Oppenheim, and especially its part concerning immigrants, reveals these same facts. Oppenheim was mainly concerned with solving the problem of population density on the route of the Baghdad and Hejaz railways, by referring to solutions already implemented in America. As we already know, the Germans desired to boost the potential fertility along the Baghdad railway, while the main difficulty was low population. Oppenheim believed that one way to increase the population was to transfer emigrants who had to leave the lost lands of the empire to the region. He suggested to transfer migrants from the Balkans, Crete, North Africa and Malaysia along the railway lines. If the first transfer were properly made, it would trigger other transfers in the future.[140]

Hajjis could be a potential solution to the housing problem along the Hejaz railway. Oppenheim offered to settle Muslim hajjis coming from all over the world along the railway lines, with the help of immigration offices. The caliph could promote this process. He also highlighted the role that *Aşiret Mektebi* (the Tribal School) could play in encouraging Bedouins to settle. Properly educated, Bedouin sheiks' children could be integrated with the system.[141]

The Ottoman government did try to put Oppenheim's suggestion into practice. In a session of the Komisyon-ı Ali on July 18, 1904, two related decisions were taken: for the accommodation of Muslim immigrants, new villages would be built along the line where food and water were abundant, and tribes living close to the line would be encouraged to work in the construction of new railroads.[142] Some families enduring economic hardships, having heard that some Muslim immigrants would be provided with accommodation, lands, livestock, seeds and tools, began to apply to government offices. For instance, Halil bin Hasan from Cyprus applied with a request to reside in Abdal Alnı District of Adana. The Muslim Refugees Commission replied that the related area was not available at that moment but, if desired, they could settle on any fertile and suitable land along the Hejaz railway.[143] Likewise, it was reported that an area 25 meters wide from the tracks was to be renovated for settlement.[144]

During the Russo–Ottoman war, Muslim Circassian immigrants fleeing from Russia began to settle in many parts of the empire, including Syria. It was reported that 80–200 Circassian families had settled there and each family was granted ten hectares of land.[145] They were exempt from military service and tax for a few years. The English Consul at Damascus, Dickson, reported that in 1888 Circassian immigrants in Jawlan district boosted grain production and trade by harvesting in the plain.[146]

As a result, new settlements such as Amman and Zarqa were created with several thousand Circassian residents. For instance, Amman was an area with ancient Greek and Roman ruins. After Circassian immigrants settled in the area, a new and vibrant town was born. Similarly, three new residential areas were built with the help of Circassians in 1903. They harvested the land and increased the number of trees.[147] In return for the privileges they had granted, the Ottoman government employed Circassian immigrants for the protection of the Hejaz railway. Shakib Arslan reported that when the British attacked the region in 1918, the one and only German troop was largely supported by Circassians.[148] The Circassian case stood as a unique example, though. The new settlement plans could not be implemented; thus the planned growth could not be achieved. A report by the accountant of the Syrian Foundation stated that if the sultan had helped settle more Circassian and Rumelian immigrants who took shelter in the Ottoman Empire along the Hejaz railway, he would have received their blessings and prayers. It is clear that their potential was not fully mobilized.[149]

The Hejaz railway could not boost industrial production either. In Syria and Hejaz regions, the level of industrial activity was low, which did not change dramatically after the railway began to operate. In fact, we may even maintain that with the import of cheap European goods via the railway, some local artisan shops in the region had to downsize.

It might have been possible to make use of the Es-Salt phosphate layers and Hama sulphur mines. German geologist Max Blanckenhorn, whom we have occasionally referred to in this study, discovered phosphate layers in the northeast of Es-Salt, which were 2.5 km long, 1 km wide and 4.5 km deep. One of the layers promised 100–150 tons of phosphate annually.[150] Aspiring to be a prime mover in this sector, Andrew Hunter & Co from London and Schilmann & Bene from Hamburg applied to the Ottoman government. Annual phosphate consumption in Germany was 200,000 tons and was expected to double in the near future. Given that 1 ton of this substance was worth 80 marks, potential players began competiting against one another. Later on, Husayn Haydar from Damascus contacted Izzat Pasha on the issue. Together with representatives from both British and German companies, investment negotiations were launched.[151]

Another reason why the French Damascus–Hama Railway Company endeavoured to gain a concession on a possible railway to be built between Jaffa and Amman was to seize the chance to excavate the Es-Salt phosphate layers.[152] However, Abdülhamid granted the concession to excavate and process these metals to the Hejaz railway Administration. Although the administration delegated this power to Nazif al-Khalidi and his French partner Viktor Azram for 40 years, they have yet to achieve any result.[153] If the project of phosphate generation had been realized, the Amman–Es-Salt line would have extended to Afula and been linked to the Haifa–Dera branch line.[154] This would have rendered the planned line more attractive for tourists and for the import of grapes and raisins.[155] The attempts to excavate phosphate continued even after the collapse of the Ottoman Empire. In a news report dated November 14, 1967, King Husayn of Jordan referred to a similar railway project to decrease the cost of transport.[156] Interestingly, another article on the same subject mentioned this "railway tradition" and referred to the aspirations of Wilhelm II and Abdülhamid II.[157]

Nevertheless, it was still believed that the Hejaz railway would be able to boost external tourism. Numerous ancient ruins in the area triggered such expectations. The Hejaz Railway Administration tried to capitalize on this potential. For example, Maan station was 6 hours away from the ancient city of Petra, so the administrators opened a small hotel in Maan and planned a main road between the two cities.[158] In *Deutsche Levante Zeitung*, a tourist group was reported as having visited Petra. In the article, the role the Hejaz railway played in tourism was stressed, and the foundation of a tourist camp near the city by a tourism company called Cook und Son was mentioned.[159] However, we should note that the expectations to revitalize tourism in the area were not fulfilled. The city lacked the required infrastructure, and the efforts to build one were not enough. The more than 40 degree heat did not help either.

Overall, the Hejaz railway made no remarkable difference in the economic and social life of the Hejaz. The railway's only real contribution was the influence of the Haifa–Dera–Damascus branch line on the life of Syria and Palestine. We already know that social and economic life in Haifa became livelier after the Hejaz railway. As we have mentioned in chapter four, in Haifa, the vibrant economic life proved advantageous primarily for the Germans. A small port town with a population of 4,000 people in 1868, it turned into an import and export hub in 1911, with a population of 23,000. A school was founded for army officers and workers employed on the railway; foreign post offices were opened, and the embassy became more active than ever. Even a branch office for the Ottoman Bank was launched. In contrast to all those positive developments, the union established by railway workers under the title *Teavün Cemiyeti* could not survive against the centralization policies of the constitutional monarchy and was closed down when the law banning trade unions was enacted (the *Tatil-i Eşgal* law).[160]

In addition to Haifa, although to a lesser extent, economic and social life also improved in a small residential area near the railway to the north of Maan. Amman, Dera, Zarqa, Hauran and Karak now had the chance to market wheat and other agricultural products. For instance, wheat transport from Hauran to Haifa doubled from 1903 to 1910.[161] However, owing to the reasons mentioned above, new arable lands were not opened either.

Among the cities that developed remarkably thanks to the Hejaz railway was Maan in the southeast. It was a small town of 3,000 people when it first became Ottoman territory. The economic and social life were positively affected by the railway. Ottoman authorities made serious attempts in this respect. For instance, a report by the Syrian governor's office emphasized that the region was highly critical in terms of agriculture and trade, and it was decided that "... a road be built from Maan to the district of Vadi-yi Musa where there are some historical monuments, and from there to the Dead Sea, and a hotel be built. If the government could not undertake the construction, the project could be subcontracted to an Ottoman company."[162]

Maan was socio-economically developed not only because of the railway but also thanks to the Ottoman garrison and the telegraph wire. With the arrival of the garrison, law and order was secured, blood feuds diminished considerably, the population increased, merchants from Palestine and Damascus began to visit the place bringing fruit, vegetables, coffee, sugar and clothes. In a short time, the price of the goods brought via railway decreased. Meanwhile, caravan trade continued; for instance, rice was transported from Aqaba by camel.[163] People from nearby towns came to Maan for jobs.[164] Postal services improved, as well. Officials and army officers used the postal service to receive letters and newspapers.[165] In addition to Maan, positive developments in Tabuk and other places where Circassians settled have already been explained in previous chapters.

As previously discussed, economic factors did not play a role in the decision to construct the Hejaz line. The socio-economic growth triggered by the Haifa–Dera–Damascus railway in Palestine and Syria was not repeated in the Hejaz. There were some districts in Palestine and Syria which partially developed, but that was thanks to the Haifa–Dera branch line. As we know, the main motivation to build that line was to deliver tools imported for the Hejaz railway to the construction site. More importantly, the DHP company had offered a very high price for transportation of these and imposed some restrictions.[166] Unlike the Hejaz, the building of a railway in this region created some dynamism, since there was already a tremendous potential for industrial growth. This very dynamism encouraged Ottoman authorities to introduce some micro and macro changes. For instance, the governor of Beirut talked about his satisfaction with the rate of industrial and agricultural growth due to the railway, emphasizing the importance of new branch lines, postal and telegram services to connect nearby towns. Some of the suggestions at a macroeconomic level included elimination of export taxes, introduction of new regulations to facilitate customs procedures, and using the same currency across the empire.[167]

All in all, the Hejaz district did not enjoy a rapid or remarkable growth. The Haifa–Dera–Damascus branch line was 280 km long, while the Hejaz railway was projected to be 2,000 km and turned out to be 1,464 km in reality. That means the former was one-seventh of the Hejaz railway as it existed on paper, and one-fifth of the one actually built. In spite of this, the Haifa–Dera–Damascus line accounted for three-fourths of the entire Hejaz railway in terms of transportation traffic. In other words, the remaining four-fifths of the railway (1,184 km) accounted for only one-fourth of the transportation of goods, material and passengers. From the south of Amman to the Hejaz railway, the only vibrant time of the year was the hajj season. Therefore, if economic expectations had been determinative, then only the 280 km Haifa–Dera–Damascus railway would have been built and this would have been adequate. The fact that they attempted to build the remaining 1,700 km can only be explained by religious/political and military/political concerns. That is why, in this study, the religious, political and military factors behind the construction of the Hejaz railway have been emphasized more than the others.

CONCLUSION

Ottoman railways emerged as an outcome of the interaction between internal and external forces. Since economic production did not reach a level that necessitated railways, the internal dynamic was determined more by military/administrative concerns, with financial/ economic factors playing a secondary role. The European efforts to capitalize on the agricultural potential and market opportunities of the Ottoman Empire constituted the fundamental external dynamic. Railway construction became the most effective instrument for the imperialist countries that were vying to acquire areas of influence within the Ottoman territory.

All Ottoman railways, except for the Hejaz railway, were built and operated by foreign capital. The empire did not have the necessary capital accumulation to finance these railways, nor the technical expertise to operate them. This alone demonstrates the uniqueness of the Hejaz railway in the history of Ottoman railways. Despite his predecessor Abdulaziz's failed attempt, Abdülhamid embarked on the construction of the Hejaz railway and mobilized the required funding. This book has clearly shown that financial and technical problems cannot account for the failure to complete the Hejaz railway.

The imperial firman (*irade-i seniyye*) that Abdülhamid issued on May 1, 1900, concerning the Hejaz railway was initially considered far-fetched and unconvincing. The reason the project sounded so ambitious was that it was to be built and operated entirely by the

Ottoman Empire. The "sacred line" would be a pure Muslim construct. Foreign capital would not be raised and it would be built by Muslim engineers only.

Taking into account the circumstances of the empire on the eve of the 20 century, it was quite understandable that the project should be regarded as unrealistic. What were the factors that motivated Abdülhamid to initiate such an ambitious project at a time when most of the budget was allocated to settlement of debts, and when salary payments were sometimes suspended? Generally speaking, it is economic factors which come to the fore in any railway construction project. The Hejaz railway, however, may be considered one in which economic factors played hardly any role at all. Dr Toncic, the Austrian consul to Jeddah, described the Hejaz as the poorest land in the world, rife with plague, cholera, squalor and banditry. As a matter of fact, Hejaz province was an economic burden on the shoulders of the empire.

The Hejaz railway project was based entirely on religious, military and administrative grounds. The secession of 5.5 million Christian subjects as a result of the loss of territories in the Balkans had transformed the empire into an overwhelmingly Asian and Muslim state. According to Abdülhamid, this change of population made it impossible to cling to liberal/Ottomanist policies, which rested on confidence towards the west and relied on the Christian elements of the empire. Only Islam could be a connecting force defining Ottoman identity. The new policy was thus a pragmatic one; it legitimated autocratic rule, exploited Muslim symbols and high-lighted the Muslim identity of the empire to retain the imperial lands where Muslims lived. Relations with Muslim countries grew ever closer, clerics were sent to Islamic countries to promote the office of the caliph, pro-Islam newspapers were subsidized, and closer ties were forged with religious orders all with the aim of making Yıldız Palace the Vatican of the Islamic world.

Europe interpreted Abdülhamid's burgeoning relations with the Muslims of the world as pan-Islamism and was intimidated by his policy. However Abdülhamid, a ruler who was fully aware of the limits of his power and thus pursued a policy of maintaining the status quo, would obviously not try to rally all Muslims of the world to his political leadership. So what did the caliph intend to do? Von der Goltz Pasha, who apparently analyzed Abdülhamid quite well, claimed that the caliph sought "to conquer from within". By boosting his reputation and hence strengthening the office of the caliph, the sultan was trying to keep the Muslim elements of the empire together. Put differently, the purpose was to prevent the virus of nationalism from infecting non-Turkish Muslims.

In the aftermath of the Ottoman–Russian War of 1877–78, Britain gradually abandoned its policy of protecting the territorial integrity of the Ottoman Empire and began instead to assert control over strategic locations on the road to India. It was evident that Britain would start playing the Arab card in line with its new foreign policy. Arab newspapers were campaigning for an Arab caliph; leaflets were distributed in cities calling people to dispense with Ottoman tutelage. A British newspaper, the *Standard*, wrote openly that Britain, with 56 million Muslim subjects, should rule the holy cities, as well.

It was clearly critical for Abdülhamid, who was trying to pursue a caliphate-centered Islamist policy, to retain Mecca and Medina. The sultan, intent on preventing the expansion of the Arab nationalist/ secessionist movement, countered by allocating a larger share of the state budget to Arabia and recruiting more Arabs to the state administration. Abdülhamid was trying to further this policy by establishing good relations with leaders of renowned Arab families and Bedouin sheiks. *Aşiret Mektebi* (The Tribal School) was established to educate the children of Arab notables and integrate them with the Ottoman system.

The function the Hejaz railway fulfilled should be seen in this light: "the sacred line" was part and parcel of the Arabia policy that the sultan was trying to implement on this slippery ground. Construction of the Hejaz railway could have consolidated the position of the sultan/caliph in the eyes of Ottomans and of the Muslim world. Given that 95,000 hajjis from India and Iran visited the holy cities every year and one-fifth of them died during the arduous journey, it is obvious why the Hejaz railway project was welcomed so enthusiastically by the Islamic world. The Ottoman project was covered by the press of India, Egypt and other Muslim countries and devout Muslims were called upon to help the caliph in his endeavor. Abdülhamid garnered extraordinary support from the Islamic world and boosted his reputation immensely. A columnist for the magazine *Die katholischen Missionen* summarized the situation as follows: "[He] became the most popular man and the hero of the Islamic world all at once."

The sustainability of this quickly-earned reputation depended first and foremost on his ability to maintain authority over the holy cities. But Britain's gaining control of the Suez Canal weakened the empire's hand. If Britain decided to close the Canal, it would disconnect the Ottoman Empire from the region. The construction of the Hejaz railway, on the other hand, would allow the empire to dispatch troops to Syria and the Hejaz without needing to rely on the Suez Canal. This would in turn strengthen Ottoman authority in the region, enable the empire to rein in the Bedouin tribes and prevent them from plundering the pilgrim caravans. Furthermore, the independence of the amirs of Mecca would be curtailed, paving the way for administrative and military reform in Arabia. Abdülhamid thus brought forward the Hejaz railway project even prior to the conclusion of the negotiations for the Baghdad railway. In fact, for Abdülhamid, the Baghdad and Hejaz railways constituted parts of a single whole. That was why he declared it his heartfelt wish to see the railway reach Mecca from İstanbul.

Contrary to expectations, the Ottoman administration was able to raise the required funds for the Hejaz railway. Abdülhamid took an important step by calling upon devout Muslims both inside and outside the empire to donate to the project. In his appeal, the caliph emphasized the fact that the Mecca line, far from a mere Ottoman railway, was a joint effort of the entire Islamic world. Since the goal was to facilitate the pilgrimage and end the suffering of devout Muslims on the way to Mecca, all Muslims, Ottoman or not, ought to contribute to the financing of the line.

The donation campaigns outside the empire were organized mostly in countries where the empire had diplomatic missions. The most interesting of these campaigns was in India. The Muslims running the campaign in this country established a Hejaz Railway Central Committee, with headquarters in Hyderabad and offices in many cities. Ottoman consuls, the local Islamist press, heads of religious orders, Muslim clerics, local notables and merchants were the most important advocates of the campaign. The British press, on the other hand, published articles against the Hejaz railway: in an attempt to dissuade Indian Muslims from donating, they claimed that contributions would be used to finance Abdülhamid's personal expenditures rather than the railway project. A ban on wearing Hejaz railway medals, which were given to all those who financially contributed to the railway construction, was the most concrete evidence of the growing British opposition in India.

The donations collected in Egypt for the Hejaz railway concerned the British even more, since Egypt was legally an Ottoman territory. The campaign in Egypt was run using the same methods as those in India. The Egyptian newspapers supported the campaign with fervent articles. Also Muslims from Morocco, Russia, China, Java and Sumatra etc. donated to the Hejaz railway. Nevertheless, the external donations constituted only 9.5 per cent of total donations and were not at the desired level. Loans issued by Ziraat Bankası constituted 12.2 per cent; taxes and other compulsory contributions represented the lion's share—41.5 per cent of total revenues. The equity of Ziraat Bankası comprises taxes as well; therefore, if the bank's loans are considered to be a form of tax, more than half of the investment for the Hejaz railway can be said to have come from compulsory contributions. Donations, most of which were collected within the empire, ranked second, accounting for 28.4 per cent of all revenues. Adding the income generated through the sale of the hide of slaughtered animals, this percentage increases to approximately 34 per cent. Apparently, one-third of the Hejaz railway project was financed through donations, something which is quite unprecedented in economic history. At the end of the day, Ottoman society financed almost the entire project through donations, loans issued by Ziraat Bankası, taxes and other compulsory endowments.

After the construction of the Hejaz railway had started, it soon became obvious that it would proceed more slowly than anticipated. The most important reason for this was the inexperience of Ottoman engineers in railway construction. The *Hendese-i Mülkiye*, which trained the first engineers of the empire, had been established only 16 years earlier. Consequently, the Ottoman administration was forced to retreat from the principle of recruiting only local people, and started hiring foreigners as well. First, the engineers working for the Anatolian and Baghdad railways were transferred to the Hejaz railway. The technical team was led by an experienced German named Meissner. Remarkable progress was observed in the construction after this restructuring.

Abdülhamid was intent on recruiting Muslim engineers for the Hejaz railway despite their inexperience. In 1904, 17 Muslim engineers were hired at his request. In order to increase this number, Ottoman engineers were sent to Germany every year to improve their technical skills. Over time, Muslim Ottoman engineers started working at all levels of the Hejaz railway project. For example, only Muslims were employed at the southern part of al-Ula, where Christians were denied entrance. The section between al-Ula and Medina was successfully completed by Muslim engineers and Muslim staff under Haji Muhtar Beg.

Construction of the Hejaz railway was plagued by a variety of problems. The shortage of water was one of the biggest issues. Particularly in the south, the problem became unbearable. Water was carried by tailor-made traincars and camels. As the construction moved southwards into the desert areas, cities and villages were few and far between, making it almost impossible to find workers. Consequently, privates of the Fifth Army deployed in Syria were recruited to work on the construction. Provision of board for more than 7,000 soldiers was an issue on its own.

Ottoman rulers had to abandon the policy of using locally-produced material on the railway. There was an effort to produce rails out of scrap iron in *Tersane-i Amire* (the Imperial Shipyard); however, rail that were manufactured were not hard enough to function properly. Similarly, the traverses made from trees could not withstand the tremendous heat in Arabia. Eventually, all railway equipment was imported from other countries, particularly from Germany. The handsome profit that German capitalists enjoyed as a result of their close relationship with the Ottoman administration has been discussed in the relevant part of this book.

The imported material was initially transported by the Beirut–Damascus–Muzeirib railway operated by DHP. This not only increased transportation costs tremendously, but also gave the French company leverage to block the process. The construction of an extension line from Dera on the Hejaz railway to Haifa on the Mediterranean cost was seen as a panacea for the problem. The Haifa–Dera line was opened in October 1905.

Meanwhile the negotiations with DHP continued. If the Damascus–Muzeirib line could have been purchased from the French, the construction of the Hejaz railway would have commenced from Muzeirib instead of Damascus and saved them from building the initial 120 km. Although the Ottoman administration was ready to pay an amount very close to DHP's bid, French Foreign Minister Delcasse objected to returning the railway concession on the grounds that it would harm French interests and reputation in Syria. For the foreign minister, French influence in the region mattered more than the company's bottom line, revealing once again the role of railway concessions in securing areas of influence. Failure to reach a consensus resulted in the construction of the Damascus–Muzeirib line, the initial section of the Hejaz railway, parallel to the French railway.

After the completion of the Haifa–Dera railway, the Ottomans started competing against the French DHP. They gained a competitive edge by dramatically cutting transportation fees. However, when the Ottoman Empire again found itself short of funds, they had to make a number of concessions in order to persuade the French to lend money. One of the preconditions that the French imposed on the loan agreement was to set a floor for transportation fees on the Haifa line. The other conditions were so harsh that, as Cemal Pasha would later admit, "in order to conclude the deal, the French imposed conditions that went as far as claiming our lives". The French would no longer permit the Hejaz Railway Authority to construct any railway through Syria and Palestine, which they regarded as their area of influence. The Ottoman administration, fearing that French pressure might go so far as to demand that the Haifa–Dera branch line be sold to them, transferred the ownership of the entire Hejaz railway to the Ministry of Foundations. According to the rules of Islam the railway, having assumed the status of a foundation (*waqf*), could not be sold.

In 1905, an extension line was planned between Maan and Aqaba. This would not only solve the problem of transportation—which became more and more expensive in the south—but also enable rapid deployment of soldiers to the region without resort to the Suez Canal. Strong British resistance to the construction of this line was to be expected, since the Aqaba railway would improve the prospects for a possible military operation in Egypt. A significant confrontation was averted only because Abdülhamid recognized the seriousness of the British objections and retreated from his plan. The Aqaba branch line was never built. The Aqaba dispute was instructive in that it revealed the limits of British tolerance towards the construction of the Hejaz railway.

The Germans were closely watching the French and British tactics to impede the Hejaz railway and seeking ways to outflank them. The Germans' wholehearted support for the project is quite under-standable. It was quite obvious that the Germans would capitalize on their rapprochement with the Ottoman administration and use the Hejaz railway to further increase their influence in Syria, where the French and British had historically had the upper hand. German influence had already expanded from Anatolia to Arabia thanks to the Anatolian/Baghdad railways. The Hejaz railway created an opportunity for the Germans to gain new areas of influence. Abdülhamid considered the Baghdad and Hejaz lines parts of a single railway project; once he abandoned the principle of not employing foreign personnel on the Hejaz railway, all prominent positions were occupied by German engineers. This not only fueled demand for German industry, but also paved the way for the creation and strengthening of German colonies in some regions. For instance, when the Haifa line was opened, the economic recovery in Haifa redounded to the advantage of the German colony living in the city, as explained in great detail in a report titled, "German Interests in the Hejaz Railway", prepared by Hardegg, the German consul to Haifa. Hardegg's self-righteousness in justifying the need for German involvement in the fight against French designs on the Hejaz railway clearly revealed the complexity of the imperialist struggle within the Ottoman Empire. Hardegg highlighted "the French desire to hamper the development of Haifa, where Germans dominated, and instead to promote Beirut, which was under French influence". The ambassador Wangenheim's remarks clarified the German approach towards the Hejaz railway even further: "Even though one might think it is no concern of ours, we have to keep a close eye on the French demands. For these demands constitute a threat against the Hejaz railway which we believe is in our best interest and hence [to be] supported extensively henceforth." This attitude, which laid claim to the Hejaz railway as if it were a German initiative, was not unique to diplomats. The German press espoused a similar line of thought. In a series of articles published in *Schlesische Zeitung,* "the frenchification of the Hejaz railway, to which the Germans made tremendous contributions" was presented as unacceptable, a development to be fought against. It was also suggested that the French experience be emulated and strenuous efforts be made in opening German schools and disseminating the German language in the Holy Land.

In addition to opposing French demands in Syria, the Germans tried to counter the British efforts to block the Aqaba line, as well. Two famous German authors can be cited in this regard. According to Max von Oppenheim, Britain was striving to sever the holy cities and the caliphate from the Ottoman administration. Therefore they were trying to encircle Mecca and Medina from three sides. The first two were Yemen in the south and Kuwait in the northeast. The third location was Aqaba in the northwest, which explains the importance they attached to the Aqaba line. All three locations were traditional pilgrimage routes. Oppenheim gave considerable weight to the merging of the Baghdad and Hejaz railways following the completion of the Aqaba line. If this were achieved, the British threat could be averted altogether. Oppenheim pointed to the prevalence of pan-Islamism in Egypt, India and other Muslim colonies and argued that Britain was mainly afraid that the caliph would fire up the spirit of insurgence among the Muslim colonies and impinge on British rule. Oppenheim argued that a "less cowardly" Ottoman sultan could have turned the tide in the Islamic world against the British. As long as this potential threat was on the table, Britain had to avoid a war in Europe, particularly an attack on Germany. Thus the German emperor had to be on friendly terms with the sultan. Apparently, Oppenheim was aiming to steer the caliph/sultan towards a truly pan-Islamist policy. We know, however, that Abdülhamid was aware of his limitations and opted for a prudent policy instead. Although the German provocation did not work on Abdülhamid, it played a pivotal role in shaping the pan-Islamist ideal of the Union and Progress movement.

Paul Rohrbach, underlining the political and military significance of the Hejaz railway, explained the importance of the Aqaba line for Germany: "One could deal a deadly blow against Britain only at one spot. This is Egypt. The conquest of Egypt by a Muslim state like Turkey will threaten Britain's influence on sixty million Muslims in India. ..." As these and similar examples show, German administrators and ideologues took the pan-Islamist project more seriously than did the Ottomans themselves.

Despite all the challenges, the lines had reached Medina in 1908; 1,464 kilometers of the Hejaz railway had been completed and opened. The next step was to construct the Medina–Mecca and Jeddah–Mecca lines as promised to the Islamic world. But this time, the strong resistance of the amirs of Mecca and the Bedouins forestalled the construction. The primary reason behind the Bedouin resistance was economic. They earned significant amounts of money by renting camels to pilgrims and providing for their needs, including food. Besides, the Bedouins received from the state an annual allowance called *urban surra* in return for not attacking the caravans. The tribes ran the risk of losing their income if the rails reached their territory. Furthermore, Bedouins were aware of the fact that the Hejaz railway would strengthen Ottoman rule in the region.

A stronger state authority in the region would mean new obligations, such as military service and tax, and eventually put an end to their waywardness. Therefore, as the Hejaz line approached Medina, Bedouin resistance grew stronger.

The Bedouins set aside their own internal disputes and collaborated for the first time in their history in order to prevent the construction of the new lines. They persistently damaged rails, dismantled traverses, and murdered soldiers. Finally the Ottoman administration had to concede defeat: the construction was aborted and the *urban surra* resumed.

The amirs of Mecca were backing the Bedouins covertly, although ostensibly they sided with the Ottoman Empire. The amirs, who had held political power for centuries, had strengthened their position even further through the end of the 19th century. Thanks to their influence on the Bedouins, they had become a natural partner in sharing the camel transport yields. Moreover, Ottoman sultans would send an annual payment known as *atiye-i humayun* and transfer half of the customs revenues of Jeddah to the amirs. The amirs were thus the staunchest defenders of the status quo. They did their best to impede the Hejaz railway, which threatened to restore Ottoman authority in the region at their expense.

Another frustration for the Ottomans was the French obstacle to the construction of the Afula–Jerusalem railway. It had been evident back in 1908 that the line to be extended to Jerusalem from the Afula juncture on the Haifa–Dera railway would bring with it considerable advantages. Moreover, the cost of construction would be very low. Nevertheless, only 17 km of the planned railway were built. The reason behind the failure to continue was the French influence in the region. The Afula–Jerusalem line would compete against the Jaffa–Jerusalem railway, which was operated by the French. The Ottoman administration had to give in to French pressure because of their desperate need to sign a loan contract with the French government. Among the conditions imposed by France were a series of taxes, railway and port concessions and termination of the Afula–Jerusalem railway construction. The Ottoman efforts to build railways on their own territory ended in stalemate. The British pressure in Aqaba, the

French influence in Syria and Palestine, the resistance of the amirs and Bedouins backed by the British around Medina, all precluded the Ottoman administration's continuing with the Hejaz railway despite its best intentions.

Under these circumstances it becomes very difficult to claim that Abdülhamid was able to accomplish the religious/political ends associated with the Hejaz railway. Even excluding the 150,000 or so pilgrims who reached Mecca by passing through the Arabian desert, only 10 per cent of the remaining travelers used the Hejaz railway. 90 per cent of the pilgrims preferred to travel by sea since the Medina–Mecca section of the Hejaz railway was never built. Those who arrived at Medina had to endure an additional 450 km of journey to reach Mecca. Obviously, the pilgrims did not wish to travel by camel under the blazing sun and threat of Bedouin attack. Unlike those who set out from Syria, they opted for a sea voyage to Jeddah, then covered the remaining 74 km to Mecca by camel. If this short line of 74 km had been built, at least those who traveled by sea would have taken the train from Jeddah to Mecca.

The military/political expectations for the Hejaz railway were also not fully realized because of the failure to complete the Medina–Mecca and Jeddah–Mecca lines. Nevertheless, the railway proved militarily useful once the Damascus–Maan line was inaugurated in 1904. For instance, the insurgence in Yemen in 1905 was suppressed by 28 military units deployed rapidly from Syria. The troops covered the distance between Damascus and Maan—previously a journey of 12 days—in 24 hours. After this point, however, the troops had to survive an arduous journey. The 120 km journey from Maan to Aqaba on the Red Sea could be completed in four days, which demonstrated the potential value of the Aqaba railway once again. Another interesting example of the improved security in the region concerns the inhabitants of Tabuk, who had run away from Bedouin attacks but decided to return to their homeland once the railway was opened.

The Hejaz railway resulted in the expansion of some settlements—particularly those in Syria—and bolstered Ottoman authority in these locations. The villages around Dera, Maan and Amman and the semi-settled Bedouins were significantly affected by this development. A stronger Ottoman domination meant military service, tax liability, surrendering arms to the Ottoman administration and similar obligations. The inhabitants of the region were not accustomed to such subordination. Insurrections were not uncommon. For instance, in 1910 a serious insurgency broke out in Karak, which is 30 km from the Hejaz line. Around 3,000 inhabitants rioted against Ottoman rule for the reasons enumerated above. Following this, the Bedouins living in the vicinity attacked the Qatrana station, destroyed the building and murdered the personnel. Still, the riot was quelled relatively quickly thanks to troops deployed by railway. It should, however, be stated that similar attacks against the stations continued for years.

Briefly it can be concluded that the inauguration of the Hejaz railway strengthened Ottoman authority in Syria to a certain extent. The Ottoman government could not, however, increase its political power in the Hejaz as much as it did in Syria. Ottoman influence could not penetrate outside the three big cities and the railway line. The desert areas far from the railway were completely under the control of the Bedouins.

To appreciate the military importance of the Hejaz railway, one needs to look at its function during World War I. With the onset of the war, the French railways in the region were handed over to the Ministry of War and new lines with military importance began to be built. The most important construction was the Hejaz–Egypt railway. By the end of 1915, the Hejaz–Egypt line had started to contribute to military deployment to Beer Sheva, 25 km away from the Canal. A series of defeats by the British, however, prevented the continuation of the lines to the Suez Canal. When in 1916 Cemal Pasha's second offensive against the Suez failed and was met with a British counterattack, the Ottoman administration had to remove the lines it had taken such pains to construct in order to prevent their being seized by the British.

Tremendous effort was expended in order to keep the lines running and benefit from their military potential during the war. Fuel supply for the trains constituted a problem. Since the entire Syrian coast was under the control of the Allied powers, it was not possible to ship coal to the region. Consequently, trains started running on wood instead of coal. Towards the end of the war, once the section of the Baghdad railway between the Taurus and Amanos mountains was built, coal coming from Germany by the Baghdad and Hejaz railways could be used. Another important predicament was the growing shortage of spare parts. Due to the unavailability of spare parts, the workshop in Damascus fell idle and the number of working locomotives decreased considerably.

When Sharif Husayn decided to enter the war on the side of Britain, a new front had to be opened in the Hejaz, which would require partial relocation of Ottoman armaments. From that time on, the already formidable cost of keeping the railway operational soared. Medina, the last Ottoman stronghold in the Hejaz, could sustain itself only thanks to the Hejaz railway. Against all odds, the Ottomans were able to retain Medina until the end of the war. However, many sources, including the memoirs of Lawrence and Ali Fuat Erden, indicate that the Ottomans were tricked by the British into trying to defend Medina at all costs. Moreover, since the Suez Canal was closed, the only way to dispatch soldiers and ammunition to the Palestinian and Sinai fronts was to keep the Hejaz lines operational no matter what. Yet as Erden states, the deployment to Hejaz was to the detriment of the Palestinian front. The strength of the latter depended on the frailty of the former. As the governor of Hejaz, Vehib Beg, said, it would perhaps have been a better idea not to open the Hejaz front at all.

In the final analysis, the Hejaz railway played a crucial role during World War I. Nonetheless, the inability to complete the sections as initially planned undermined its military potential. First of all, it was not possible to deploy troops to the region from İstanbul by a non-stop journey. The Hama–Aleppo section connecting the Hejaz railway to the Baghdad railway had been completed in 1907, but the Baghdad line was interrupted twice—by the Taurus and Amanos mountains—making transportation very difficult. This connection was built only towards the end of the war. The third stopover was in Rayak, where the lines narrowed. Besides, the British had already prevented the construction of the Maan–Aqaba line. The Aqaba line was key, for it would provide an exit to the Red Sea. We have already discussed the challenges faced by Cemal Pasha during the Suez operation because of the failure to construct the Aqaba line.

The Hejaz railway did not revitalize the economic life of the region. Only the Haifa–Dera branch line brought about some positive developments in a handful of settlements in Syria and Palestine. It created remarkable dynamism in the economic and social life in Haifa. A small port town which had been inhabited by 4,000 people in 1868, Haifa became an export and import hub with a population of 23,000 in 1911. Although not comparable with the rate of development in Haifa, some settlements close to the railway in the north of Maan benefited from the railway to a certain extent. For instance, the amount of wheat sent from Hauran to Haifa doubled from 1903 to 1910. However, the Hejaz did not enjoy similar growth. The traditional economic and social structure in Hejaz province continued to prevail. The developments around Syria and Palestine were positive externalities of the Haifa–Dera branch line, rather than an outcome of deliberate initiatives.

The most concrete indicator of the fact that the Hejaz railway was not built with the expectation of economic benefit is the Haifa–Dera–Damascus line which, though constituting one-fifth of the Hejaz railway, accounted for three-fourths of the passenger and freight traffic. As a matter of fact, apart from the pilgrimage season, there was no activity on the railway starting from the south of Amman. If economic expectations had been decisive, it would have sufficed to build the 280 km Haifa–Dera–Damascus line. The construction of the additional 1,700 km can be explained only by religious/political and military/political factors.

NOTES

Introduction

1. A poet expressing his feelings about railways. Herbert Heaton, **Economic History of Europe**, New York, 1948, pp. 506, 507.
2. Eric J. Hobsbawm, **The Age of Revolution 1789–1848**, New York, 1964, p. 63.
3. Op. cit., p. 202.
4. Murat Özyüksel, **Osmanlı İmparatorluğu'nda Nüfuz Mücadelesi, Anadolu ve Bağdat Demiryolları**, İstanbul, 2013.

Chapter 1   The Historical Development of Railway Construction in the Ottoman Empire

1. Chesney would complain years later that his efforts for "this transportation project which would lead to India" were not appreciated by prominent decision-makers in Britain. Francis Rawdow Chesney, **Narrative of the Euphrates Expedition**, London, 1868, p. viii.
2. Muhteşem Kaynak, "Osmanlı Ekonomisinin Dünya Ekonomisine Eklemlenme Sürecinde Osmanlı Demiryollarına Bir Bakış" **Yapıt**, no. 5, 1984, p. 68.
3. G. D. Clayton, **Britain and the Eastern Question: Missolonghi to Gallipoli**, London, 1971, p. 91.

4. Engin Deniz Akarlı, **Belgelerle Tanzimat, Osmanlı Sadrazamlarından Ali ve Fuat Paşaların Siyasi Vasiyetnameleri**, İstanbul, 1978, p. 25

5. Tithe in this case is an Ottoman tax assessed as a proportion of the revenues accruing from a particular piece of tilled land.

6. Vedat Eldem, **Osmanlı İmparatorluğu'nun İktisadi Şartları Hakkında Bir Tetkik**, Ankara, 1970, p. 153.

7. Orhan Kurmuş, **Emperyalizmin Türkiye'ye Girişi**, İstanbul, 1977, p. 57.

8. For detailed information regarding the Treaty of 1838, see Mübahat S. Kütükoğlu, **Osmanlı-İngiliz İktisadi Münasebetleri II**, (1838–1850), İstanbul, 1976, pp. 7–42 et al. Also valuable are: Seyfettin Gürsel, "1838 Ticaret Antlaşması Üzerine", **Yapıt**, no. 10 (Nisan-Mayıs 1985), pp .27–36 and in the same issue Taner Timur, "1838'de Türkiye, İngiltere Rusya", pp. 4–26.

9. For detailed information about the challenges faced (e.g. camels could not walk through water courses; the government confiscated them when deemed necessary), see Ali Akyıldız, "Osmanlı Anadolusunda İlk Demiryolu: İzmir-Aydın Hattı (1856–1866)", in **ÇYO**, ed. E. İhsanoğlu, M. Kaçar, İstanbul, 1995, p. 250.

10. Kurmuş, pp. 58, 59.

11. Hermann Schmidt, **Das Eisenbahnwesen in der asiatischen Türkei**, Berlin, 1914, p. 3

12. MacDonald Stephenson, **Railways in Turkey**, London, 1859, p. 7. Akyıldız, p. 252.

13. Kurmuş, pp. 74, 75.

14. Charles Morawitz, **Türkiye Maliyesi**, Maliye Bakanlığı Tetkik Kurulu Yayını, no: 1978–188, p. 324.

15. M. Hecker, "Die Eisenbahnen der asiatischen Türkei", **AfEW**, Berlin, 1914, p. 780.

16. A. du Velay, **Türkiye Maliye Tarihi**, Maliye Bakanlığı Tetkik Kurulu Neşriyatı no. 178–1978, Ankara, 1978, pp. 386, 387; Morawitz, p. 325.

17. Şevket Pamuk, "Osmanli Imparatorluğu'nda Yabancı Sermaye: Sektörlere ve Sermayeyi İhraç Eden Ülkelere Göre Dağılımı (1854–1914)", **ODTÜ Gelişme Dergisi**, 1978 Special Edition, p. 148.

18. Yaqub N. Karkar, **Railway Development in the Otoman Empire 1856–1914**, New York, 1972, p. 79.

19. For instance, Submission of Saffet Pasha, B.O.A., Y. EE, 44/146.

20. Kurmuş, pp. 88–98, 242, 248 et al.

21. Mihail P. Guboğlu, "Osmanlı İmparatorluğu'nda Karadeniz Tuna Kanalı Projeleri (1836–1876) ve Boğazköy - Köstence Arasında İlk Demiryolu İnşası (1855–1860)", in **ÇYO...**, p. 239.

22. "The Sultan's First Railway Journey", The **Times**, 27.8.1867, p. 9.

23. "The newest and fastest of trains rushed the astonished princes from Dover to London in a little over two hours", Joan Haslip, **The Sultan, The Life of Abd ul Hamid II**, London, 1958, p. 52.

24. N. Heintze, "Eisenbahnen in der Türkei", **Mitteilungen der geographischen Gesellschaft**, vol. XI, Hamburg, 1896, p. 45; Paul Dehn, **Deutschland und die Orientbahnen**, München, 1883, p. 14.

25. Vahdettin Engin, **Rumeli Demiryolları**, İstanbul, 1993, pp. 62, 63.

26. Op cit., pp. 90, 91.

27. George Hallgarten, **Imperialismus vor 1914**, München, 1963, Band 1, pp. 228–45. For Hirsch's behind the scenes manipulations that led to losses for the Ottoman treasury, see **B.O.A., Y. PRK. TNF**, 4/39 and 2/35; **Y.PRK.ML**, 7/38. Ambassador Marschall likewise stated that "Hirsch systematically robbed Turkey". From Marschall to Hohenlohe, 5.3.1898, Constantinople, (confidential), **G.D.D.**, vol. II, XII 559, p. 467. Another interesting example concerning this issue can be cited from the memoirs of German Professor Ernst Hirsch, a faculty member at the İstanbul School of Law. Professor Hirsch expressed his opinion that the Baron had defrauded the Ottoman Empire, saying "... It all boils down to the cunningness of an old namesake of mine, who lived years ago: Belgian Baron Hirsch. During the years I spent in Turkey, they would ask me if we were related. In 1869 this Baron Hirsch gained the concession and obtained a kilometer guarantee as well. Thus, the rails were laid unnecessarily long, which cannot be objectively justified. Briefly stated, our train was moving on a straight plane, drawing thousands of semi-circles and, contrary to my experience of the Central European system, stopping at very small stations needlessly." Ernst E. Hirsch, **Anılarım; Kayzer Dönemi, Weimar Cumhuriyeti, Atatürk Ülkesi**, Ankara, 1997, p. 195.

28. Cemil Öztürk, "Tanzimat Devrinde Bir Devletçilik Teşebbüsü: Haydarpaşa - İzmit Demiryolu", in **ÇYO**, pp. 279–82.

29. **B.O.A., A. MKT. MHM**, 447/22.

30. Friedrich Kochwasser, "Der Bau der Bagdadbahn und die deutsche Orientpolitik", **Deutsch-Türkische Gesellschaft E.V., Bonn Mitteilungen**, (Juni, 1975), p. 1; A. D. Noviçev, **Osmanlı İmparatorluğu'nun Yarı Sömürgeleşmesi**, Ankara, 1979, p. 23

31. At the end of this borrowing process, the government issued a notification dated October 10, 1875, and had to reduce the installments and the interest rate of the internal and external debts by half for five years; see İ. Hakkı Yeniay, **Yeni Osmanlı Borçları Tarihi**, İstanbul, 1964, p. 53. This overtly meant the bankruptcy of the state; see Carl Anton Schaefer, **Ziele und Wege für die jungtürkische Wirtschaftspolitik**, Karlsruhe, 1913, p. 42

32. Mukavelename (Aggrement), **B.O.A., Y.A. Hus.**, 220/17.

33. Celal Dinçer, "Osmanlı Vezirlerinden Hasan Fehmi Paşa'nın Anadolu'nun Bayındırlık İşlerine Dair Hazırladığı Layiha", **Türk Tarih Kurumu Belgeler Dergisi**, vol. V-VIII (offprint from no. 9–12), Ankara, 1972, p. 160.

34. W. J. Mommsen, **Europaeische Finanzimperialismus vor 1914. Ein Beitrag zu einer pluralistichen Theorie des Imperialismus**, Göttingen, 1979, p. 127.

35. Donald C. Blaisdell, **European Financial Control in the Ottoman Empire**, New York, 1929, p. 128. Helmut Mejcher, "Die Bagdadbahn als Instrument deutschen wirtschaftlichen Einflusses im Osmanischen Reich", **Geschichte und Gesellschaft**, no. 1–4 (1975), p. 465.

36. Johann Manzenreiter, **Die Bagdadbahn, Als Beispiel für die Entstehung des Finanzimperialismus in Europa** (1872–1903), Bochum, 1982, p. 39. For an example of applications in the Ottoman Archives see Strousberg's application, **B.O.A., Y.A. Hus.**, 168/53. British entrepreneur Staniforth wrote a letter to Abdülhamid in order to obtain a concession for a railway which would connect İstanbul with the Persian Gulf, **B.O.A. Y.PRK.TKM**, 7/36. For the Euphrates railway project proposed by the famous British businessman William Andrew, **B.O.A. Y. EE. KP**, 2/147, for the application of Williams, an American businessman, **B.O.A., Y.PRK. A.**, 4/64.

37. İlber Ortaylı, **İkinci Abdülhamit Döneminde Osmanlı İmparatorluğu'nda Alman Nüfuzu**, Ankara, 1981, pp. 18, 43.

38. However, these British initiatives do not mean an utter deviation from their conventional Ottoman politics. Edward Grey would still define their policy towards the empire as "prevention of its collapse and dismemberment". An absolute collapse would lead to insurmountable problems such as "who takes what" and make a Europe-wide war inevitable: Elie Kedourie, **England and the Middle East, The Destruction of the Ottoman Empire 1914–1921**, London, 1987, pp. 10, 11.

39. "Amongst the big states, the British are the most unreliable... In July, 1882 Admiral Seymour had announced that they did not plan to invade Egypt... However renegade Albion (Britain) forgot his promises very quickly", Sultan Abdülhamit, **Siyasi Hatıratım**, İstanbul, 1984, pp. 127, 128.

40. For the details of the rapprochement in military relations that began with the German officers sent upon Abdülhamid's request, Özyüksel, pp. 49–51. Jehuda Wallach, **Anatomie einer Millitaerhilfe**, Düsseldorf, 1976, pp. 36–61.

41. Kurt Zander, "Einwirkungen der kleinasiatischen Eisenbahnen auf die Hebung des Grundbesitzes", **AfEW**, Berlin, 1894, p. 942

42. Blaisdell, p. 133.

43. Karl Helfferich, **Georg von Siemens, Ein Lebensbild aus Deutschlands Grosser Zeit**, Berlin, 1923, Band III, pp. 44, 50.

44. Dr. Jaeck asserts that the foundations of the Baghdad railway were laid during this trip, Jaeck, "Die Beziehungen der deutschen Industrie zum türkischen Reiche", **Technik und Wirtschaft**, no. 5, May 1916, pp. 193, 194. For Wilhelm's trip, also see C. A. Engelbrechten, **Kaiser Wilhelms Orientreise und deren Bedeutung für den deutschen Handel**, Berlin 1890. See also Joan Haslip, **II. Abdülhamid**, İstanbul, 1998, pp. 209–24

45. Wilhelm's note was "good"; from Radowitz to Caprivi, 4.7.1890, **G.P.**, Band XIV/ II, nr. 3598, pp. 441, 442.

46. Helfferich, Band III, p. 63. For Chambon'a threats, **B.O.A., Y. PRK.EŞA.**, 16/78 and **B.O.A., Y.A. Hus., 268/105.**

47. From Radolin to Marschall, **G.P,** Band XIV/II, nr. 3970, p. 456. For a similar attitude by the British Embassy, **B.O.A., Y.A. Hus.,** 268/124.

48. Max Schlagintweit, **Reise in Kleinasien,** München, 1898, p. 29.

49. In addition to all the factors discussed above, Weltpolitik should be interpreted as a mechanism of defense against domestic demands for socialism and democracy by winning glorious victories in foreign policy. See John Lowe, **The Great Powers, Imperialism and the German Problem, 1865–1925,** London and New York, 1981, pp. 141–6. The viewpoint was expressed mostly by historians of the Fischer school. For Fischer's approach, Fritz Fischer, **Griff nach der Weltmacht,** Düsseldorf, 1964, pp. 15–36.

50. Lothar Rathmann, **Stossrichtung Nahost 1914–1918,** Berlin, 1963, pp. 25, 26.

51. For the details of the report, Özyüksel, pp. 125–8.

52. Jan Stephan Richter, **Die Orientreise Kaiser Wilhelms II 1898,** Hamburg, 1996, p. 89.

53. A. S. Jerussalimski, **Die Aussenpolitik und Diplomatie des deutschen Imperialismus Ende des 19. Jahrhunderts,** Berlin, 1953, p. 682.

54. For the details of its implementation, **B.O.A., Y. PRK. TNF,** 6/73. For Damat Mahmut Pasha's support, **B.O.A., Y. EE,** 84/42; **B.O.A. Y. PRK. TKM,** 16/35; Sina Akşin, **Jön Türkler ve İttihat ve Terakki,** İstanbul 1980, pp. 37, 38.

55. Helfferich, vol. III, pp. 102, 103, 105.

56. C. Mühlmann, "Die deutschen Bahnunternehmungen in der asiatischen Türkei", **Weltwirtschaftliches Archiv,** vol. 24, Jena, 1926, p. 374.

57. Deutsche Levante Linie and Norddeutscher Lloyd wanted to engage in maritime activities. See Lothar Rathmann, "Zur Legende vom antikolonialen Charakter der Bagdadbahnpolitik in der wilhelminischen Aera des deutschen Monopolkapitalismus", **Zeitschrift für Geschichtswissenschaft,** Sonderheft IX (1961), p. 249.

58. Bagdad Railway Aggrement, **B.O.A., Y.PRK.TNF,** 7/44 or **Y. EE,** 1/24. **Bagdad Railway, Convention of March 5, 1903,** Published by His Majesty's Stationery Office, London, 1911, p. 42.

59. "Eine Konzession der anatolischen Bahnen auf Ausbeutung der Petroleumquellen in Mesopotamien", **Frankfurter Zeitung,** 12.9.1904, PA/AA, Türkei 152. Also see Minister of Hazine-i Hassa Ohannes Paşa, **B.O.A., Y. MTV,** 260/15. See also Ortaylı, pp. 92, 93.

60. Richard Hennig Friedenau, **Die deutschen Bahnbauten in der Türkei,** Leipzig, 1915, p. 8; Hüber p. 45

61. Anton Mohr, **Der Kampf um Türkisch Asien, Die Bagdadbahn,** Meissen (n.a.), p. 51.

62. For the German attempts in the Adana region, see Paul Uhlig, **Deutsche Arbeit in Kleinasien von 1883 bis 1918,** Greifswald, 1925, pp. 140–8;

Lothar Rathmann, **Berlin-Bağdat, Alman Emperyalizminin Türkiye'ye Girişi**, İstanbul, 1982, p. 109; Rıfat Önsoy, **Türk-Alman İktisadi Münasebetleri (1871–1914)**, İstanbul, 1982, pp. 55, 56; Ortaylı, p. 92; Imbert p. 43. Regarding the irrigation of the Konya Plain see "Die Bewaesserung der Ebene von Konia", **Korrespondenzblatt der Nachrichtenstelle für den Orient**, Sonderbeilage zu Nu. 4 des III. Jahrganges, 1916, p. 153; Willy Reibel, **Die Gründung auslaendischer Eisenbahnunterrehmungen durch deutsche Banken**, Düsseldorf, 1934, pp. 47, 48; Rosa Luxemburg, "Emperyalizmin Mısır ve Osmanlı İmparatorluğu'na Girişi", in Rathmann, **Berlin-Bağdat...**, p. 185.

63. Mühlmann, p. 379; Reibel, p. 68; Mohr, p. 51; Schmidt, p. 70.
64. Şehmus Güzel, "Anadolu - Bağdat Demiryolu Grevi", **Tanzimat'tan Cumhuriyet'e Türkiye Ansiklopedisi**, vol. III, (1985), p. 830. Dr Arhangelos, the chief of railway health services, stated that "One of the flowers that blossomed thanks to constitutional monarchy is the Trade Union of Government Officials and Workers". See Gavriel Arhangelos, **Anadolu Osmanlı Demiryolu ve Bağdat Demiryolu Şirket-i Osmaniye İdaresi'nin İçyüzü**, İstanbul, 1327, p. 236. Donald Quataert, **Social Disintegration and Popular Resistance in the Ottoman Empire, 1881–1908: Reactions to European Economic Penetration**, New York and London, 1983, pp. 71–93.
65. **Meclisi Mebusan Zabıt Ceridesi**, 35. Session, vol. 2, Ankara, 1982, pp. 62–76. Gündüz Ökçün, "Osmanlı Meclis-i Mebusanında Bağdat Demiryolu İmtiyazı Üzerine Yapılan Tartışmalar", A.Ü.S.B.F., XXV, 2 (Haz. 1970), pp. 22–43. Regarding the fact that the Anatolian peasants covered the total cost of the railway construction through the taxes that they paid, Rosa Luxemburg, Karl Liebknecht, Franz Mehring, **The Crisis in the German Social Democracy**, New York, 1818, pp. 41–65.
66. "... However the French in Paris stipulated conditions which stood for forcing the Turkish Economy under French control and no government that has self-esteem can concede", Feroz Ahmad, **İttihatçılıktan Kemalizme**, İstanbul, 1985, p. 46.
67. Earle Edward Mead, **Turkey, The Great Powers and the Bagdad Railway, A Study in Imperialism**, New York, 1923, p. 225.
68. Karl Helfferich, **Die deutsche Türkenpolitik**, Berlin, 1921, p. 23.
69. B.O.A., HR. SYS, 110/11. Also, Max Wiedemann, **Bagdad und Teheran: Politische Betrachtungen und Berichte**, Berlin, 1911, p. 41.
70. From Kiderlen to Hollweg, 24 Mai 1911, **G. P.**, 27/2, nr. 10038, p. 688.
71. Mejcher, p. 477; Mühmann, p. 384; Rathmann, **Berlin....**, p. 102
72. Haluk Ülman, **1860–1861 Suriye Buhranı, Osmanlı Diplomasisinden Bir Örnek Olay**, Ankara, 1966, pp. 5–9.
73. Şehâbeddin Tekindağ, "Lübnan", **İslâm Ansiklopedisi**, vol. 7, p. 105.
74. Ülman, pp. 36–42, 48, 49.
75. İlber Ortaylı, "19. Yüzyıl Sonunda Suriye ve Lübnan Üzerinde Bazı Notlar", **Osmanlı Araştırmaları**, vol. 4, İstanbul, 1984, p. 97.

76. Adolf Beer, **Allgemeine Geschichte des Welthandels im neunzehnten Jahrhundert**, Band 2, Wien, 1884, p. 12.
77. Ülman, pp. 79–81, 89, 115.
78. M.S. Anderson, **The Eastern Question 1774–1923, A Study in International Relations**, New York, 1966, p. 157.
79. Ulrich Fiedler, **Der Bedeutungswandel der Hedschasbahn; Eine historisch geographische Untersuchung**, Berlin, 1984, p. 14.
80. Hecker, p. 798.
81. **Wiener Politische Correspondenz**, 25 May 1900.
82. Martin Hartmann, "Das Bahnnetz Mittelsyriens", **Zeitschrift des Deutschen Palaestina Vereins**, Band 17, (1894), p. 62.
83. From the Tarabya Embassy to Prime Minister Schillingsfürst, 23.5.1900, **PA/AA**, Türkei 152.
84. Jacob M. Landau, **The Hejaz Railway and the Muslim Pilgrimage, A case of Ottoman Political Propaganda**, Detroit, 1971, p. 11.
85. **ZdVDEV**, 2.5.1908.
86. Fiedler, pp. 19, 20.
87. From Hardegg to Hollweg, 11.8.1913, Haifa, **PA/AA**, Türkei 152.
88. Hecker, p. 797. Reinhard Hüber, "Die Entwicklung der Eisenbahnen in der europaeischen und asiatischen Türkei, insbesondere in Syrien", **ZdVDEV**, nr. 34, 29.4.1908, p. 564. The idea to construct a railway in Jerusalem was first introduced in 1838 by a British Jew, Moses Montefiore. Also see Werner Sölch, **Expresszüge im Vorderen Orient**, Düsseldorf, 1989, p. 35.
89. Fiedler, p. 13.
90. A. Ruppin, **Syrien als Wirtschaftsgebiet**, Berlin, 1917, p. 308; Hecker, p. 797.
91. Hermann Guthe, **Die Hedschasbahn von Damaskus nach Medina, ihr Bau und ihre Bedeutung**, Leipzig, 1917, p. 3.
92. Fiedler, p. 21. Every day four trains, each with a speed of 16 km per hour, traveled both ways, Hecker, p. 1310.
93. From Högging to Berchtold, 31.3.1914, Berlin, **HHStA**, Admin. Reg., F 19/33.
94. Feza Günergün, "Osmanlı Devleti'nde Buharlı Tramvay İşletme Teşebbüsleri", in **ÇYO**, pp. 385, 386.
95. Martin Hartmann, "Das Bahnetz Mittelsyriens", **Zeitschrift des Deutschen Palaestina-Vereins**, Band 17. (1894), p. 57; Hecker, p. 782.
96. **B.O.A., Y.A.Hus.**, 411/171.
97. Fiedler, pp. 16–18.
98. Note titled "Sefirin İfadatı", which mentions the argument made by the French, 19 May 1890, **B.O.A., Y. E.E.**, 5/13.
99. Note signed by Rıdvan, 20 April 1891, **B.O.A., Y. PRK. AZJ**, 16/74.
100. Süreyya Pasha, 27 May 1890, **B.O.A., Y. E.E.**, 5/13.
101. **P.D.H.C.**, fifth series, vol. 59, p. 2169.
102. **P.D.H.C.**, vol. 59, p. 2188.

103. **Wiener Politische Correspondenz,** 5.7.1901.
104. From Micksche to Calice, 11.6.1893, Beyrut, **HHStA,** Botschaftsarchiv Konstantinopel, K. 80.
105. **B.O.A., Y.A.Hus.,** 411/171. On the same issue, from Marschall to Schillingsfürst, 26.2.1899, Pera, **PA/AA,** Türkei 152.
106. Grand Vizier Rıfat Paşa, **B.O.A., Y.A. Hus.,** 411/171.
107. **Hamburgischer Correspondent,** 8.2.1900.
108. Grand Vizier Rıfat Pasha, **B.O.A. Y.A.Hus.,** 410/27. For further information about the British Embassy's failure to prevent the concession from being cancelled, from Marschall to Bülow, 28.11.1902, Pera, **PA/AA,** Türkei 152.
109. Hans Jürgen Philipp, "Der bedunische Widerstand gegen die Hedschusbahn", **Die Welt des Islams,** vol. XXV, (1985), p. 34. Also see From Richard to Schillingsfürst, 17.8.1897, Bagdad, **PA/AA,** Türkei 165.
110. Philipp, p. 34. From Richarz to Schillingsfürst, 11.2.1897, Bagdad, **PA/AA,** Türkei 165. A British Military Intelligence Service correspondence dated October, 9 1898, stated that in time of war the railway would be easily destroyed by local tribes, who could be bribed by the enemy (Turk, French or Russian). Orhan Koloğlu, **Abdülhamit Gerçeği,** İstanbul, 1987, p. 321.
111. Richard Hennig (Friedenau), "Das Projekt der transarabischen Bahn", **DLZ,** 15.10.1912, Nr. 17, pp. 4–6.
112. The Austrian Consul had drawn attention to the unfeasibility of the Kuwait–Jeddah railway projects: "Such an initiative would never be welcomed by Turkey and the Amir of Nejd would also oppose it", from Toncic to Berchtold, 20.3.1913, Jeddah, **HHStA,** Admin. Reg., F 19/33.
113. From Hardegg to Hollweg, 23.6.1913, Hayfa, **PA/AA,** Türkei 152.
114. Rifat Uçarol, **Gazi Ahmet Muhtar Pasha (1839–1919), Askeri ve Siyasi Hayatı,** İstanbul, 1989, pp. 196, 244.
115. Ufuk Gülsoy, **Hicaz Demiryolu,** İstanbul, 1994, pp. 38, 39. Muhtar Pasha would raise those opinions frequently in the future. As an example, **B.O.A., Y.A.Hus.,** 482/160
116. Fritz Lorch, "Die Eisenbahn Jaffa-Port Said", **DLZ,** no. 11, 15.7.1912, p. 17.

Chapter 2   Decision to Construct the Hejaz Railway

1. Martin Hartmann, "Die Mekkabahn, ihre Aussichten und ihre Bedeutung für den Islam", **Asien,** Nr. 10, (1912), p. 148
2. From Marschall to the Foreign Ministry, 3.5.1900 and 7.5.1900, Pera, **PA/AA,** Türkei 152.
3. Komisyon-ı Ali's decision stressing the importance of the issue. 29.4.1908, **B.O.A., MV,** 118/99.
4. "Die Mekkabahn", **Die Reform, Internationales Verkehrsorgan,** 4. Jahrgang., 23. Heft (Erstes Augustheft 1903), pp. 1420, 1421.

236   NOTES TO PAGES 40–44

5. From Marschall to the Foreign Ministry, 3.5.1900, Pera, **PA/AA**, Türkei 152.
6. From Marschall to Schillingsfürst, 5.5.1900, Pera, **PA/AA**, Türkei 152.
7. From Wangenheim to Schillingsfürst, 10.7.1900, Tarabya, **PA/AA**, Türkei 152.
8. İhsan Süreyya Sırma, "Fransa'nın Kuzey Afrika'daki Sömürgeciliğine Karşı Sultan II. Abdülhamid'in Panislâmist Faaliyetlerine Ait Bir Kaç Vesika", **İstanbul Üniversitesi Edebiyat Fakültesi Tarih Enstitüsü Dergisi**, 7–8, (1977), p. 178.
9. From Calice to Gotuchorski, 9.5.1900, İstanbul, **HHStA**, Admin. Reg., F 19/18.
10. Enver Ziya Karal, **Osmanlı Tarihi**, vol. 8, Ankara 1988, p. 471
11. İsmail Hakkı Uzunçarşılı, **Mekke-i Mükerreme Emirleri**, Ankara, 1972, p. 23
12. Münir Atalar, **Osmanlı Devletinde Surre-i Hümâyun ve Surre Alayları**, Ankara, 1991, pp, 93–136. Uzunçarşılı, p. 36 et al. Also see Mehmet Zeki Pakalın," Surre Alayı", **Osmanlı Tarih Deyimleri ve Terimleri Sözlüğü**, vol. III, İstanbul 1993, pp. 280–3.
13. Uzunçarşılı, p. 41
14. Atalar, p. 207. Uzunçarşılı, pp. 42, 59
15. From Toncic to Aehrenthal, 30.3.1909, Jeddah, **HHStA**, Admin. Reg. F 19/18.
16. From Marschall to Hohenlohe, 5.5.1900, Pera, **PA/AA**, Türkei 152.
17. Fiedler, p. 40.
18. Sultan Abdülhamit, **Siyasi Hatıratım**, İstanbul, 1984, p. 145.
19. Bayram Kodaman, **Sultan II. Abdülhamid Devri Doğu Anadolu Politikası**, Ankara, 1987, p. 82. On the same topic Georgeon wrote, "The ratio of Muslims to the total population of the empire increased from 68% to 76% in a couple of years. As of now, three quarters of the empire are Muslims." See François Georgeon, "Son Canlanış (1878–1908)", in **Osmanlı İmparatorluğu Tarihi II**, ed. Robert Mantran, İstanbul, 1995, p. 148.
20. Andre Raymond, "Arap Siyasal Sınırları İçinde Osmanlı Mirası", in **İmparatorluk Mirası**, ed. L. Carl Brown, İstanbul, 2000, p. 164.
21. Hasan Kayalı, **Jön Türkler ve Araplar**, İstanbul, 1998, p. 34.
22. Haluk Ülman, **Birinci Dünya Savaşı'na Giden Yol**, Ankara, 1973, p. 144.
23. Kayalı, p. 37. Goltz Pasha even suggested to shift the Ottoman capital to Asia Minor in order to exert equal influence on the two fundamental components of Ottoman society, op. cit., p. 36. In the coming years, von der Goltz would suggest a Turkish–Arab Empire modeled on the Austro-Hungarian Empire.
24. For detailed information see Cezmi Eraslan, **II. Abdülhamid ve İslâm Birliği**, İstanbul, 1992, pp. 217–27.
25. For instance, Ahmed Cevdet Pasha and Lütfi Effendi had interpreted the *Tanzimat* as a tactical move to enlist help from the West against Mehmed Ali Pasha. See Taner Timur, "Osmanlı ve Batılılaşma", in **Osmanlı Çalışmaları**, Ankara, 1989, p. 86.

26. Murat Özyüksel "İktisadi Davranışlarımızın Tarihsel Kökenleri Üzerine Bir Deneme" İktisat Dergisi, no. 380, June/July 1998, p. 168.

27. For an interesting analysis of how Abdülhamid and "classical" right-wing parties of the Republican era formulated their policies with backing from this front see İdris Küçükömer, Düzenin Yabancılaşması, İstanbul, 1969, particularly p. 58 and after.

28. Engin Akarlı, "II. Abülhamid (1876–1909)", Tanzimattan Cumhuriyete Türkiye Ansiklopedisi, vol. 5, İstanbul, 1985, p. 1293.

29. Şerif Mardin, stated that Abdülhamid pursued his policy of Islam via "agents working in Central Asia, North Africa and the Far East", see Şerif Mardin, "İslamcılık", Türkiye'de Din ve Siyaset, İstanbul, 1991, p. 16.

30. For instance, until 1910 sermons were preached in the name of Abdülhamid in Zanzibar. İlber Ortaylı, "19. Yüzyılda Panislamizm ve Osmanlı Hilafeti", Türkiye Günlüğü, no. 31 (November – December 1994), p. 27.

31. Chedo Mijatovich's remarks are interesting to note: " Abd ul Hamid did not pursue the panislamist movement; panislamist movement reached out and found him ", see Chedo Mijatovich, "Abd ul Hamid", Die Zukunft, no. 47, (22.8.1908), p. 296

32. In the official meetings made by Russian and English diplomats, from time to time, pan-Islamism was mentioned as an important common threat. In the examples mentioned below, it was argued that also Ittihadists had a tendency to display a pan-Islamist ideology as a realpolitik, B.D., vol. X, Part I, pp. 512, 567, 583, 598, 600, 601, 622, 623, 630 and vol. X, Part II, pp. 2, 3.

33. "For Islamic countries rapidly becoming colonies, the Ottoman Empire was their last resort which could help them restore their tarnished honor." See Mümtaz'er Türköne, Siyasi İdeoloji Olarak İslamcılığın Doğuşu, İstanbul, 1991, p. 171.

34. Kemâl H. Karpat, "Pan-İslamizm ve İkinci Abdülhamid: Yanlış Bir Görüşün Düzeltilmesi", Türk Dünyasını Araştırmaları Dergisi, no. 47 (June 1987), p. 27.

35. Sırma, "Fransa'nın Kuzey Afrika'daki...", pp. 159 and 162, 163.

36. Kološlu, p. 200–12. Karpat, pp. 13, 14, 28. In Maalouf's novel Semerkant, Jamal ad-Din al-Afghani says the following: "Am I not your guest ? Let me go! If I am your prisoner then fetter me and put me in the dungeon!", Amin Maalouf, Semerkant, İstanbul, 1998, p. 152. For an analysis of how Afghani rejected Abdülhamid's conception of Ottomanism and pan-Islamism see Bessam Tibi, Arap Milliyetçiliği, İstanbul 1998, p. 118.

37. Max von Oppenheim's report, 22.6.1907, Berlin, PA/AA, Türkei 152.

38. Peter Hopkirk, On secret Service East of Constantinople: The Plot to Bring Down the British Empire, London, 1994. This book is entirely about Germany's efforts to attract militant Islamic forces to its cause with the help of the Ottoman empire. For Wilhelm Wassmus's efforts in Afghanistan, who is dubbed as the Lawrence of Germany, see pp. 63, 87, 122 et al.

39. "Muslims around Aleppo were bombarded with such an intense propaganda that they were made to believe that Kaiser was a Muslim and that Germans were fighting against Russia in the name of Islam." German and Turkish propagandists regarded Kaiser as "Haji Wilhelm, friend and guardian of Islam". See Philip H. Stoddard, **Teskilat-ı Mahsusa**, Istanbul, 1993, p. 59; pp. 15, 16, 20, 31, 93, 94.

40. **Neue Freie Presse**, 3.9.1908. Goltz Pasha had expressed similar opinions in his foreword to Auler Pasha'a book. See Auler Pasha, **Die Hedschasbahn, Auf Grund einer Besichtigungsreise und nach amtlichen Quellen**, Gotha, 1906, von der Goltz's Foreword, p. 2. Apparently Auler Pasha shared the same opinion, op. cit., p. 62

41. From Bodman to von Bülow, 21.8.1904, Tarabya, **PA/AA**, Türkei 152.

42. Selim Deringil, "Osmanlı İmparatorluğu'nda 'Geleneğin İcadı', 'Muhayyel Cemaat' ('Tasarımlanmış Topluluk') ve Panislamizm", **Toplum ve Bilim**, no. 54/55, (Summer/Autumn 1991), pp. 52, 62.

43. "Sultan ... knew that the British ambassador was watching the delegations coming from Afghanistan and Turkmenistan to İstanbul with anxious and fearful eyes", Haslip, pp. 139 and 202.

44. Cevdet Ergül, **II. Abdülhamid'in Doğu Politikası ve Hamidiye Alayları**, İzmir, 1997, p. 37.

45. "Die Hedschasbahn, der Islam und Englands Stellung dazu", **Die Grenzboten**, Nr. 38, 17.9.1908, p. 573.

46. The British dominance in Basra was underpinned by their secret agreement with the Mubarak al Sabah, Amir of Kuwait. According to this agreement, sheik had conceded British control in return for money. Historical Section of the Foreign Office, **Persian Gulf**, vol. XIII, London, 1920, p. 52. For the text of the agreement, G. U. Aitchison, B.G.S., **A Collection of Treaties, Engagements and Sanads, Relating to India and Neighbouring Countries**, Delhi, 1933, pp. 262–7. Britain had signed similar agreements with the sheiks of Qatar, Bahrain and Oman, Tahsin Paşa, **Tahsin Paşa'nın Yıldız Hatıraları**, İstanbul, 1990, p. 349.

47. İhsan Süreyya Sırma, **Osmanlı Devleti'nin Yıkılışında Yemen İsyanları**, İstanbul, 1994, pp. 76–78 et al.

48. Abdülhamid was fully aware of the situation: "Unfortunately Britain has a very strong influence in Arabia. They have already started causing trouble in Yemen. Arab tribes provoked by the British are rioting one after the other." Sultan Abdülhamit, pp. 144, 145.

49. Koloğlu, Abdülhamit..., p. 175. Eraslan, p. 268.

50. Georgeon, "Son Canlanış...", p. 160. Georgeon, **Sultan Abdülhamid**, İstanbul, 2006, p. 225.

51. Sultan Abdülhamit, p. 144.

52. Georgeon, "II. Abdülhamid ve İslam", **Tarih ve Toplum**, no. 112, Nisan 1993, p. 48; Kayalı, pp. 41, 43.

53. Ernest Dawn, **From Ottomanism to Arabism**, Urbana, Chicago London, 1973, p. 140. et al. This book elaborates on Kevakibi's opinions and the evolution of Arab nationalism, pp. 122–47 et al.

54. Peter Mansfield, **The Ottoman Empire and its Successors**, London and Basingstoke, 1973, p. 18.

55. Sultan Abdülhamit, p. 144.

56. Max Roloff-Breslau, **Arabien und seine Bedeutung für die Erstaerkung des Osmanenreiches**, Leipzig, 1915, p. 14.

57. Eraslan, pp. 209, 210.

58. Suraiya Faroqhi, **Hacılar ve Sultanlar (1517–1683)**, İstanbul, 1995, p. 200, 203. Also see C. Snouck Hurgronje, **Mekka**, Haag, 1889, p. 38 and after.

59. Hell, almost echoing Abdülhamid, wrote "Mecca and İstanbul. This is Islam, the history of Islam". See J. Hell, "Stambul und Mekka", **Erlanger Aufsaetze aus ernster Zeit** (1917), p. 59.

60. Azmi Özcan, **Pan - İslamizm, Osmanlı Devleti, Hindistan Müslümanları ve İngiltere (1877–1914)**, İstanbul, 1992, pp. 172, 173.

61. Kayalı, p. 45.

62. Sayyid Ahmad Khan, stated that Abdülhamid could have no claim to be the caliph of Muslims in India and that Islam did not leave room for a universal conception of caliphate anyway. See Deringil, p. 58.

63. Ortaylı, "19. Yüzyılda...", pp. 27, 28.

64. Eraslan, p. 195. For Lütfi Pasha's detailed justification of Ottoman caliphate please see Hulusi Yavuz, **Osmanlı Devleti ve İslamiyet**, İstanbul, 1991, pp. 95–110.

65. Kayalı, pp. 21, 37, 38; Georgeon, "Son Canlanış", p. 162; Karl K. Barbir, "Bellek, Miras ve Tarih: Arap Dünyasında Osmanlı Mirası", in **İmparatorluk Mirası**, p. 152.

66. Kodaman, p. 89.

67. Auler Pascha, **Die Hedschasbahn**, p. 64.; Kodaman, p. 89. For detailed information, Bayram Kodaman, "II. Abdülhamid ve Aşiret Mektebi", **Türk Kültürü Araştırmaları**, XV/1–2, (1976).

68. İbrahim Sivrikaya, "Osmanlı İmparatorluğu İdaresindeki Aşiretlerin Eğitimi ve İlk Aşiret Mektebi", **Belgelerle Türk Tarihi Dergisi**, vol. XI, no. 63, (1972), p. 17.

69. Alişan Akpınar, **Osmanlı Devletinde Aşiret Mektebi**, İstanbul, 1997, pp. 25, 27.

70. Aydın Talay, **Eserleri ve Hizmetleriyle Sultan Abdülhamid**, İstanbul, 1991, p. 155.

71. Akpınar, p. 26.

72. From Oppenheim to Bülow, 29.4.1903, Cairo, **PA/AA**, Türkei 152, p. 15. For a similar comment, Pallavicini to Aehrenthal, 6.3.1908, İstanbul, **HHStA**, Admin. Reg., F 19/18, pp. 30, 31.

73. İlber Ortaylı, **İmparatorluğun En Uzun Yüzyılı**, İstanbul, 1983, p. 108.

240 NOTES TO PAGES 55–61

74. İlber Ortaylı, "Osmanlı İmparatorluğu'nda Arap Milliyetçiliği", **Tanzimat'-tan Cumhuriyet'e Türkiye Ansiklopedisi**, vol. 4, İstanbul, 1985, p. 1035.
75. Sultan Abdülhamit, p. 145.
76. Hugo Grothe, **Meine Studienreise durch Vorderasien (Kleinasien, Mesopotamien u. Persien)**, 1906 and 1907, Halle 1908, p. 38.
77. H. von Kleist, "Die Hedjasbahn", **Asien**, no. 6 (Mart 1906), p. 84.
78. Sait Toydemir, "Hicaz Demiryolu İnşaatı Tarihinden", **Demiryollar Dergisi**, no. 275–278 (1948), p. 65.
79. Gülden Sarıyıldız, "Hicaz'da Salgın Hastalıklar ve Osmanlı Devleti'nin Aldığı Bazı Önlemler", **Tarih ve Toplum**, no. 104 (Ağustos 1992), p. 20. Also see Sarıyıldız, **Hicaz Karantina Teşkilatı 1865–1914**, Ankara, 1996, p. 23 et al.
80. R. King, "The Pilgrimage to Mecca: Some geographical and historical Aspects", **Erdkunde**, Nr. 26 (1972), p. 69.
81. Internal strikes among Bedouin tribes deteriorated the problem of security. For example, **B.O.A.,Y. MTV**, 192/173.
82. Auler, p. 24
83. Andre Raymond, **Osmanlı Döneminde Arap Kentleri**, İstanbul, 1995, p. 21.
84. Frequent wires sent the the sultan reporting safe arrival of the hajj caravans to predetermined destinations illustrate the significance of the issue. See following examples, **B.O.A., Y.A. -Hus.**, 407/84; 483/115; 414/30; 414/40.
85. Faroqhi, **Hacılar...**, pp. 6–8, 88–92, 200. Atalar, pp. 221–6. Taxes for Haremeyn contribution, **B.O.A., MV**, 121/54.
86. The article by Mustafa Kâmil published in Muhammedan. **B.O.A.,Y.PRK. TKM**, 49/ 44.
87. In his article published in *al-Manar* newspaper of Eygpt, Rashid Rida argued that the love of God and the prophet required the safeguarding of God's home in Mecca and the Prophet's tomb in Medina. Since the Hejaz railway would facilitate access to and protection of these holy places, it had to be supported by all Muslims alike. Another newspaper in Eygpt, which heralded the Hejaz railway fervently, was *al-Ahram*. See William Ochsenwald, **The Hijaz Railroad**, Virginia, 1980, p. 75. See also *Thamarat al Funun*, a similar publication in Beirut, Landau, p. 20. The contribution of the press to the Hejaz railway will be discussed in chapter three.
88. "Die neue Mekka-Bahn", **Die katholischen Missionen**, Nr. 10, (1906/1907), p. 219.
89. Sultan Abdülhamit, pp. 123, 124.
90. Franz Stuhlmann, **Der Kampf um Arabien zwischen der Türkei und England**, Hamburg, Braunschweig, Berlin, 1916, pp. 47, 48.
91. Muhtar Pasha's report. **B,O.A., Y.A. - Hus.**, 482/160; Grand Vizier Ferid Pasha's petition, **B,O.A., Y.A. - Res.**, 129/3.
92. Amy Singer, **Kadılar, Kullar, Kudüslü Köylüler**, İstanbul, 1996, p. 147.

93. Grand Vizier Ferid Pasha reported that the Bedouins had "attacked" the district of Harran, "in order to avoid more damage and harm both to the inhabitants and their belongings two cavalry squads ought to be dispatched urgently", **B.O.A., Y.A. - Hus.**, 406/71.

94. Singer, p. 148.

95. Abdülhamid, expressed his opinions about Ibn Saud as follows, "He had once envisioned himself as the mighty king of independent Arabia, however did not fathom that he would have had to submit to British yoke while trying to gain independence with the help of the British". Sultan Abdülhamit, p. 151

96. Imbert, p. 74.

97. Governor of Cezayir-i Bahr-i Sefid, Abidin Pasha, **B. O. A, Y.PRK. UM,** 75/87.

98. We will discuss this issue in chapter five. Please see the following for further discussion: Suraiya Faroqhi, **Herrscher über Mekka, Die Geschichte der Pilgerfahrt,** München, Zürich, 1990, pp. 197–202. See also Uzunçarşılı, pp. 23–30.

99. **B. O. A, Y.PRK. UM,** 54/68.

100. **B. O. A, Y.PRK.BŞK,** 78/1.

101. From Bülow to Wilhelm, 9. 28. 1901, **G.P.**, vol. XVII, Nr. 5235, pp. 405–7. The radical nationalist organization *Alldeutschen Verband* advocated Abdülhamid's approach because it was in line with German interests. See "Hedschas- und Bagdadbahn", **Alldeutsche Blaetter,** Nr. 49, 8.12.1906, p. 395.

102. Auler, p. 63.

103. Philipp, pp. 33, 36.

104. Hartmann, "Die Mekkabahn...", p. 148.

105. Ufuk Gülsoy, **Hicaz Demiryolu,** İstanbul, 1994, p. 33. For the list of healthcare measures Kaimakam Şakir proposed see Sarıyıldız, **Hicaz Karantina...,** pp. 80–4.

106. Military Commission's report on Osman Nuri Pasha's proposal, 02.17.1892, **B.O.A., Y.MTV,** 59/38.

107. For the influence of commission members on Abdülhamid's decision to carry on with the Hejaz railway project, Tahsin Pasha, p. 348.

108. Atalar, p. XIX. Gülsoy, p. 35. In 1893, Ottoman general staff also discussed the need to extend the Anatolian railway and connect Hejaz with the capital city, **B.O.A,Y.PRK.ASK,** 89/8.

109. **B.O.A, Y.PRK.BŞK,** 28/97.

110. The report by Şakir Pasha and İbrahim Pasha, 31 January 1893, **B.O.A., Y. MTV,** 74/54.

111. **B.D.,** vol.5, p. 8.

112. Tahsin Pasha, pp. 27, 28. **B.D.,** vol.5, p. 8.

113. Max Blanckenhorn, "Die Hedschaz-Bahn, Auf Grund einer Reisestudien", **Zeitschrift derGesellschaft für Erdkunde zu Berlin,** (1907), pp. 225, 226.

114. Gülsoy, p. 36.

115. For the impact of Egypt/Syria/Basra railway discussions in Britain on Inshaullah's thinking, Landau, p. 12.
116. Landau, p. 12; Ochsenwald, pp. 22, 71.
117. Syed Tanvir Wasti, "Muhammad Inshaullah and the Hijaz Railway", **Middle East Journal** (Spring 1998), p. 63.
118. Muhammad Inshaullah, **The History of the Hamidia Hedjaz Railway Project in Urdu, Arabic and English**, Lahore, 1908, pp. 2, 3.15, 16.
119. Inshaullah, pp. 3, 4.
120. Inshaullah, pp. 4, 7. Wasti, p. 64.
121. Inshaullah, p. 7.
122. Wasti, pp. 64, 65; Inshaullah, pp. 7, 8.
123. Inshaullah, pp. 8, 9; Philipp p. 34.
124. Philipp p. 35.

Chapter 3   Financing the Hejaz Railway

1. According to Gülsoy 18 per cent (p. 57), according to Ochsenwald 15 per cent (p. 59).
2. Meclis-i Vükela, 3 June 1900, **B.O.A., İ. TNF**, 1318. p. 1.
3. Declaration, 22 June 1900, **B.O.A., İ. ML**, 13 18.S.37. For the loan distribution per year, Ochsenwald, p. 80. For frequent interference of Ziraat Bankası, Komisyon-ı Ali, 18.11.1908, **B.O.A., MV**, 121/53.
4. For an example, "Hicaz Şimendiferi Hattı İanesi", **İkdam**, 7 July 1900.
5. "... or else, no one outside of the empire woud have donated for the strengthening of Ottoman domination", from Toncic to Aehrenthal, 30.3.1909, Jeddah, **HHStA**, Admin. Reg. F 19/18.
6. Haslip, **II. Abdülhamid**, p. 260 and Imbert, p. 81. For some other exaggerated comments see: "... In this way half of the Hejaz railway capital was obtained through aids until 1909", Özcan, p. 158; "Even the poorest peasents of India competed with one another in contributing to the railway, just like the Indian governers" Orhan Kologlu, "Hicaz Demiryolu (1900–1908) Amacı, Finansmanı, Sonucu (1900–1908)" in **ÇYO**, p. 309. "The Ottoman Empire had not met the costs, every single day thousands of voluntary donations flowed from the Islamic World", from Toncic to Aehranthal, 30.3.1909, Jeddah, **HHStA**, Admin. Reg., F 19/18.
7. Mekkabahn", **Wiener Politische Correspondenz**, 9.1.1902.
8. Komisyon-ı Ali, 2 February 1907, **B.O.A., Y. MTV**, 293/96.
9. İbrahim Artuk, "Hicaz Demiryolu'nun Yapılması ve Bu Münasebetle Basılan Madalyalar, **İstanbul Arkeoloji Müzeleri Yıllığı**, no. 11–12 (1964), p. 76; For the document listing the names of the donators from districts of Trabzon, Sivas, Diyarbakır and Ankara together with the type of medals they were rewarded with, see Komisyon-ı Ali, 4 April 1904, **B.O.A., Y. MTV**, 258/148.

10. İsmet Çetinyalçın, "Liyakat Madalyası", **VIII. Türk Tarih Kongresi, Kongreye Sunulan Bildiriler**, C. III, Ankara, 1983, pp. 1723, 1724. For the details about Abdülkadir Pasha's being rewarded with an order of merit with the aim of "standing as a model for others as well", **B.O.A., Y. EE. KP**, 21/2042.

11. Komisyon-ı Ali, 5.9.1903, **B.O.A., Y. MTV**, 250/78.

12. Ochsenwald, p. 69. Özcan, pp. 158, 166. Gülsoy, p. 74.

13. Inshaullah, particularly see the foreword of the book.

14. **Civil and Military Gazette**, 14.9.1904 and 18.6.1904.

15. From Richthofen to Bülow, 25.5.1908, Simla, **PA/AA**, Türkei 152.

16. In 1907, Muhammad Inshaullah had collected 380 pounds of donation and, together with 50 other Indian philanthropists, had been rewarded with a medal. See **B.O.A., Y.MTV**, 302/160.

17. Inshaullah's letter, 5.12. 1911, **B.O.A., İ. MBH**, 1330.M.3. See also Wasti, p. 66.

18. The translation of the Inshaullah's letter, 19 January 1911, **B.O.A., HR. TO**, 541/40. See also Inshaullah's letter, 5.12.1911, **B.O.A., İ. MBH**, 1330.M.3. Gülsoy, pp. 77, 79, 80.

19. While suggesting Azhari Abd al-Qayyum to donate the collected money to the Hejaz railway he had stated the pressmen were already printing the unprinted Islamic works one by one, thus there was no need to establish a printing press. İkdam, 4.10.1900.

20. Ochsenwald, pp. 69, 70.

21. İkdam, 04. 10. 1900.

22. Ochsenwald, pp. 70, 71. İkdam, 4.10.1900.

23. "The Proposed Damaskus Railway, Subscriptions from India", **Civil and Military Gazette**, 17 .5.1904.

24. Ochsenwald, pp. 70, 71.

25. Kologlu, "Hicaz...", p. 307.

26. **B.O.A., Y. PRK. TKM**, 50/59.

27. From the Minister of Manshawi Foundation to the Grand Vizier, 21.2.1914, **B.O.A., İ. MMS**, 1332.R.26/12. For another donation made by the same foundation, see the letter by the Deputy Minister of the Foundation, 30.1.1911, **İ. MMS**, 1329.S.22/10.

28. "The Proposed...", **Civil and Military Gazette**, 17 .5.1904.

29. Ochsenwald, p. 73.

30. From Richthofen to Bülow, 25.5.1908, Simla, **PA/AA**, Türkei 152.

31. "The Hedjaz Railway, A Record of Jobbery", **Times of India**, 11.3.1907.

32. According to Ochsenwald, between 17,000 and 40,000 liras (p. 73), according to Kologlu 15.500, "Hicaz...", p. 305.

33. "The Hedjaz... ", **Times of India**, 11.3.1907.

34. From Pallavicini to Aehrenthall, 6 March 1908, İstanbul, **HHStA**, Admin. Reg., F 19/18, pp. 32, 33. On the same subject, Ahmet Onur, **Türkiye Demiryolları Tarihi (1860–1953)**, p. 25.

35. From the London Embassy to Bülow, and annex in the report: "Great Turkish Railway", 22.7.1908, London, **PA/AA**, Türkei 152. See also **The Globe**, 22.7.1908.
36. Ochsenwald, pp. 73, 74. For the presentation of the related notebook concerning the medals given to Indian Muslim donators, Komisyon-ı Ali, 22.10.1905, **B.O.A., Y. MTV**, 279/138.
37. From Quadt to Bülow, 11.1.1907, Calcutta, **PA/AA**, Türkei 152.
38. Kologlu, "Hicaz...", pp. 307, 308.
39. The **Muhammedan**, 19 April 1906.
40. Özcan, pp. 169, 170.
41. Arnold J. Toynbee, 1920'lerde Türkiye, Hilafetin İlgası, İstanbul, 1998, p. 57.
42. From the Ministry of Public Works to the Ministry of Internal Affairs, 18 March 1911, **B.O.A., DH. İD**, 126/12.
43. From Richthofen to Bülow, 04.08.1903, San Stefano, **PA/AA**, Türkei 152.
44. Ochsenwald, p. 76
45. Gulsoy, p. 80. For the submission of the Ministry of Domestic Affairs on the money collected by the Egypt Donation Comission, **B.O.A., Y. MTV**, 206/124.
46. Ochsenwald, pp. 74, 75.
47. Related report from Richthofen to Bülow.
48. For the donation sent for the third time by al-Liva newspaper, and the document listing the donators' names together with the amounts they donated and the medals they were rewarded with, Komisyon-ı Ali, 4 .4. 1904, **B.O.A., Y. MTV**, 258/149.
49. For examples on the support of the Egyptian newspapers to the Hejaz railway, Ochsenwald, p. 75
50. For a part of the Egyptian donators list, **İkdam**, 03.10. 1900.
51. Kologlu, "Hicaz...", p. 307.
52. Herbert Pönicke, **Die Hedschasbahn und Bagdadbahn, erbaut von Heinrich August Meissner Pascha**, Düsseldorf, 1958, p. 3.
53. İkdam, 30.06.1902. In the meantime, forgers began to collect donations as well. "In China, a forger was reported to deceive people by collecting donations for Hejaz railway, with a large number of seals he claimed to have been given by the caliphate." For details, **B.O.A., Y, PRK. AZJ**, 51/18.
54. Gülsoy, p. 83.
55. Koloğlu, "Hicaz...", p. 305. For the donation of 3,000 rubles by the Governor of Bukhara, **B.O.A., Y. EE**, 53/95.
56. **Wolff's Telegraphisches Bureau**, no. 1053, 08.03. 1901, Berlin PA/AA, Türkei 152.
57. Kologlu, "Hicaz...", pp. 308, 309.
58. Sırma, " Ondokuzuncu Yüzyıl...", p. 186.
59. **Die Post**, 23.1.1904.
60. From the Consulate to Bülow, 22.6.1904, **PA/AA**, Cairo, Türkei 152.

61. "The Hedjaz... ", The Times of India, 11.3.1907.
62. B.D., vol. V, p. 8.
63. From Marschall to Schillingsfürst, 5.5.1900, Pera, PA/AA, Türkei 152. See also Vossische Zeitung, 1.3.1902.
64. Ahmet Ratib Pasha, 6 July 1900, B.O.A., Y. PRK. ASK, 163/2. For Hasan Pasha's report opposing these accusations, 4.10.1901, Y. PRK. ASK, 175/69.
65. For the 18-page Donation Instruction, B.O.A., Y.MTV, 203/83.
66. Sabah, 23 June 1900.
67. Internal Affairs Office of Correspondences, 27 January 1904, no. 3953–12, B.O.A., DH. MKT, 450/1. However, since the related application was described as "upon will", we may easily infer that there was no legal obligation, see Internal Affairs Office of Correspondences, 21 April 1904, no. 180–6, DH. MKT, 450/1.
68. Kölnische Zeitung, 25.7.1907
69. Uçarol, p. 429.
70. Komisyon-ı Ali, 4.4.1904, B.O.A., Y. MTV, 258/149.
71. İkdam, 7.7.1900. Ochsenwald, p. 67. Koloğlu, "Hicaz...", p. 310.
72. İkdam, 30.7.1900 and 17.7.1900.
73. Detailed list coming from Amirate of Mecca, 17.6.1901, B.O.A., Y. PRK. UM, 1319.S.29.
74. Meclis-i Vükela, 1.1.1902, B.O.A., MV, 103/48. For a warning by Grand Vizier Said Pasha on the same subject, B.O.A., DH. MKT, 572/38.
75. Gülsoy, pp. 68, 69. Ochsenwald, p. 64
76. Meclis-i Vükela, 3 June 1900, B.O.A., İ. TNF, 1318.S.1. For the "Islamhane" tickets printed in 1906 and sent to all the districts, Komisyon-ı Ali, 23. 3.1906, B.O.A., Y. MTV, 284/130.
77. Declaration, 22 June 1900, B.O.A., İ. ML, 1318/ S-37.
78. Ochsenwald, p. 66, Katipzade Ali Beg would give 110 liras for every 100 km. İkdam, 17.07.1900.
79. Ochsenwald, pp. 66, 67.
80. From İzmit Armenian Delegate to Prime Ministry, 22.6.1905, B.O.A., DH. MKT, 982/39.
81. From London Ambassador Kostaki to The Headclerk of Mabeyn-i Hümayun (Head Office of the Empire), 16.6.1902, B.O.A., Y. PRK. EŞA, 40/52.
82. Ochsenwald, p. 61.
83. H. Slemman, "Le Chemin de fer de Damas - La Mecque", Revue de l'Orient Chretien, Year 5 (1900), p. 524.
84. Ochsenwald, p. 65.
85. Internal Affairs Office of Correspondences, 22 October 1904, B.O.A., DH. MKT, 450–1.
86. Ministry of Internal Affairs, 29 August 1907, B.O.A., DH. MKT, 1196/88.
87. Internal Affairs Office of Correspondences, 21 April 1904, no. 180–6, B.O.A., DH. MKT, 450/1.

88. From Grand Vizier Said Pasha to Internal Affairs, 22.10.1902, **B.O.A., DH. MKT**, 608/5.

89. İkdam, 07.07.1900.

90. Komisyon-ı Ali, 4 March 1903, **B.O.A, Y.MTV**, 241/22. On duty as the Head of this Commission in 1908, Akif Pasha complained about the uncertainties that the post embodied. **B.O.A, Y. PRK. TNF**, 9/23.

91. Declaration, 22 June 1900, **B.O.A., İ. ML**, 1318/ S-37.

92. From Rosen to Schillingsfürst, 25.07.1900, Jerusalem, **PA/AA**, Türkei 152.

93. Ochsenwald, p. 68.

94. Gulsoy, p. 60.

95. Gülsoy, p. 61, 90.

96. **B.O.A., Y. EE. KP**, 15/1482.

97. "Auszug aus den die Hedjas-Bahn betreffenden Artikeln türkischer Zeitungen", 21.8.1904, **PA/AA**, Türkei 152.

98. Orhan Koloğlu, "Muktesid Musa'ya Göre Hicaz Demiryolu'nun Amacı", **Toplum ve Ekonomi**, no. 10 (July 1997), pp. 136–40.

99. Landau tranaslated all of Arif's manuscript titled as Kitâbu's-Sa'âdeti'n-Nâmiyeti'l-Ebediyye fi's Sikketi'l-Hâdidiyyeti'l-Hicâziyye into English. The summary we made below depends on the important sections of the English translation. See Landau, pp. 21–170.

100. Zekeriya Kurşun, "Şekib Arslan'ın Bazi Mektupları ve İttihatçılar ile İlişkileri" **İstanbul Üniversitesi Edebiyat Fakültesi Tarih Enstitüsü Dergisi**, Year 1995–1997, no. 15, pp. 603, 613, 614. For all of the letters, **B.O.A., DH. KMS**, 63/53.

101. Paschasius, "Die Hedschasbahn", **AfEW**, 1927, p. 1147. Also see from Wangenheim to Hohenlohe Schillingsfürst, 10.07.1900, Tarabya, **PA/AA**, Türkei 152.

102. Gülsoy, pp. 95, 96. Ochsenwald, p. 77. For the details of how the related cut-offs were transferred to the Ministry of Finance, **B.O.A., Y. MTV**, 247/139. For the amendment made in 1904, **B.O.A., ZB**, 588/80.

103. For the cut-offs, see **B.O.A.,MV**, 121/54. For the transfer of the surplus to the railway, **B.O.A.,, İ. TNF**, 1321.Ra.3. For another transfer of the same type, from the Minister of Military Schools to the Ministry of Internal Affairs, 16 February 1904, **B.O.A., DH. MKT**, 450–1.

104. From Sternburg to Bülow, 22.05.1901, Simla, **PA/AA**, Türkei 152. In the Assembly decision dated 30.12.1908, the fact that the donations are not obligatory is mentioned, **B.O.A., MV**, 123/8. See also **B.O.A., İ. TNF**, 1319. L. 2.

105. E. A. Ziffer, "Die mohammedanische Eisenbahn (Hedschasbahn)", **Zeitschrift des österreichischen Ingenieur-und Architekten Vereines**, no. 9 (1910), p. 135.

106. Internal Affairs Office of Correspondences, 5 February 1903, **B.O.A., DH. MKT**, 449/10.

107. B.O.A., DH. MKT, 710/9. From Ministry of Foreign Affairs to Internal Affairs, 7 January 1904, B.O.A., DH. MKT, 450–1. Also, see Ochsenwald, p. 77, 78
108. Komisyon-ı Ali, 14 July 1903, B.O.A., Y. MTV, 247/139. See also B.O.A., Y. EE. KP, 19/1848.
109. From Donations Commission to Ministry of Internal Affairs, 13 April 1903, B.O.A., DH. MKT, 450–1.
110. From Minister of Land Registry and Cadastre to Grand Vezirate, 23 July 1903, B.O.A., DH. MKT, 449/10.
111. Gulsoy, p. 91. Also see ZdVDEV, 22.7.1903 and 9.6.1906. On the same issue, from Marschall to Bülow, 10.01.1903, Pera, PA/AA, Türkei 152.
112. B.O.A., Y. PRK. MF, 4/69.
113. From Ministry of Education to Internal Affairs, 29 September 1903, B.O.A., DH. MKT, 450–1.
114. Ochsenwald, p. 78. Gülsoy, p. 99. See also Meclis-i Vükela, 5 January 1902, B.O.A., İ.TNF, 1319.L.2.
115. Komisyon-ı Ali, 25.12.1903, B.O.A., Y. MTV, 254/18.
116. Komisyon-ı Ali, 10.12.1906, B.O.A., Y. MTV, 285/155.
117. Komisyon-ı Ali, 20.12.1905, B.O.A., Y. MTV, 281/141.
118. ZdVDEV, 7.8.1907. See also B.O.A., Y. MTV, 250/78.
119. Komisyon-ı Ali, 25 December 1903, B.O.A., Y. MTV, 254/17.
120. Komisyon-ı Ali, 6.05.1902, B.O.A., Y. MTV, 229/61.
121. B.O.A., Y. MTV, 305/158.
122. Director of Internal Press, 1 October 1905, B.O.A., DH. MKT, 451/1.
123. For the proposal made by Deputy General Manager of Documents Ahmet Âtıf Bey, B.O.A., Y.A. - Hus., 411/168.
124. B.O.A., DH. EUM. MEM, 52/12. B.O.A., DH. MKT, 451/1.
125. B.O.A., Y. MTV, 295/185.
126. B.O.A., DH. MKT, 451/1. For the details on how the villagers had to pay taxes for the wood they had sold, DH. MKT, 1199/79.
127. From the Ministry of Internal Affairs to Prime Ministry, 26 October 1903, B.O.A., DH. MKT, 450/1. From Ferid Pasha to Ministry of Internal Affairs, 3 March 1903, DH. MKT, 449/10.
128. From Mutasarrıf of İzmit to Ministry of Foreign Affairs, 14 September 1903, B.O.A., DH. MKT, 450/1.
129. B.O.A., DH. İD, 52/5. See also B.O.A. MV, 138/26 and B.O.A., İ. MMS,1333.B.4.
130. B.O.A., Y. PRK. SH, 7/17.
131. Gülsoy, p. 101. With the assumption that some demands for privilege would be accepted more easily, they included promises to help Hejaz railway. For instance, in return for the privilege to have their ads issued, some applicants promised to grant the railway 6,000 kurush per year. The related amount was found inadequate, thus, the request was rejected. B.O.A., DH. MKT, 833/22.
132. B.O.A., Y. MTV, 295 /144.

133. Komisyon-ı Ali, **B.O.A., Y. MTV,** 258/91.
134. The interest of 1 per cent was replaced by that of 2.5 per cent, see note no. 5277 dated 4 November 1900, **B.O.A., Y. MRZ.d,** 9080.
135. From Ranzi to the Prime Ministry and the Ministry of Foreign Affairs, 04.10.1912 Damascus, **HHStA,** Admin. Reg., F 19/33.
136. Ochsenwald, pp. 79, 80.
137. Fiedler, p. 49.
138. From Pallavicini to Aehrenthal, 15.01.1908, İstanbul, **HHStA,** Admin. Reg., F 19/18.
139. Sultan Abdülhamit, p. 123. Pallavicini agrees on the issues of Izzat Paşa and medals, see aforementioned report.
140. Kologlu, "Hicaz... ", p. 305. Ochsenwald gives the ratio of external donations to the total donations as 8 per cent, this means that the ratio of donations coming from outside the borders to the total income will slightly be lower. See Ochsenwald, p. 69.
141. Ochsenwald, pp. 79–81. See also **B.O.A., İ. TNF,** 1324.M.5.
142. Schmidt, p. 127. See also from Pallavicini to Aehrenthal, 06.03.1908, İstanbul, **HHStA,** Admin. Reg., F 19/18. However, it should not be forgotten that the Hejaz railway had been constructed as narrow and the others as broad gauge.
143. Auler, p. 53. Mygind gives the kilometer cost as 1,600 liras which equals approximately 36,500 francs. See Mygind, **Vom Bosporus...,** p. 85
144. From Jenisch to von Bülow, 03.06.1905, Cairo, **PA/AA,** Türkei 152.

Chapter 4   The Construction of the Hejaz Railway

1. For the organization of the Hejaz railway administration, see **Hicaz Demiryolu, Hicaz Demiryolu'nun Vâridât ve Mesârifive Terâkki-i İnşaatı ile Hattın Ahvâl-i Umumiyyesi Hakkında Malumât-ı İhsâiyye ve İzahât-ı Lazımeyi Muhtevidir,** Year 5, 1330, İstanbul 1334, p. 1
2. Ochsenwald, p. 31
3. Auler, p. 26; Gülsoy, p. 109,; Ochsenwald, p. 27; Fiedler, p. 57.
4. Gülsoy, p. 107.
5. Tahsin Paşa, p. 28. Also see Max Schlagintweit, **Verkehrswege und Verkehrsprojekte in Vorderasien,** Berlin, 1906, p. 28.
6. From Pallavicini to Aehrenthal, 6.3.1908, İstanbul, **HHStA,** Admin. Reg., F 19/18.
7. From Bodman to Bülow, 21.8.1904, Tarabya, **PA/AA,** Türkei 152.
8. Auler, p. 25; Guthe, p. 15; Gülsoy, pp. 107, 108.
9. **B.O.A., Y. MTV,** 288/40.
10. **Hicaz Demiryolu'nun Vâridât...,** p. 1.
11. Pönicke, pp. 3, 4

12. Ufuk Gülsoy - William Ochsenwald, "Hicaz Demiryolu", **Diyanet İslâm Ansiklopedisi**, vol. 17, İstanbul, 1998, p. 442.
13. From Lütticke to Bülow, 20.1.1902, Berlin, **PA/AA**, Türkei 152.
14. Fiedler, pp. 53, 54.
15. From Wangenheim to Hollweg, 2.6.1913, Tarabya, **PA/AA**, Türkei 152.
16. For Ferid Pasha's submission, which includes the French allegations and the Ottoman arguement, **B.O.A., Y. A. - Hus.**, 482/140–1. On the same issue see **Y.A. - Hus.**, 471/41–1.
17. Ferid Pasha, **B.O.A., Y.A. Hus.**, 472/97–1.
18. Ferid Pasha, **B.O.A., Y.A. Hus.**, 471/125. On the same issue see From Marschall to Bülow, 21.5.1904, Tarabya, **PA/AA**, Türkei 152. Constanz followed the issue insistently, **B.O.A., Y.A. - Hus.**, 472/13–1. About the statement to the French ambassador of the impossibility for the Ottoman Empire to pay 7.5 million francs, **B.O.A., Y.A. - Hus.**, 472/98. For the negotiations to continue and points of conflict, **B.O.A., Y.A. - Hus.**, 473/50.
19. Submissions of the Grand Vizier Ferid and the Minister of Foreign Affairs İsmail Hakkı Pasha, **B.O.A., Y.A. - Hus.**, 473/50.
20. **B.O.A., Y.A. - Hus.**, 472/97. **B.O.A., Y.A. - Hus.**, 473/50.
21. Ferid Pasha, **B.O.A., Y.A. - Hus.**, 472/98. Under which conditions the line could be taken back, **B.O.A., Y.A. - Hus.**, 474/72. Again that the railway could be taken back when desired, **B.O.A., Y.A. - Hus.**, 473/50.
22. **B.O.A., Y.A. - Hus.**, 482 /140.
23. From Padel to the Ministry of Foreign Affairs, 25.1.1905, Pera, **PA/AA**, Türkei 152.
24. Ferid Pasha, **B.O.A., Y.A. - Hus.**, 483/118.
25. From Marschall to the Ministry of Foreign Affairs, 25.1.1905, Pera, **PA/AA**, Türkei 152.
26. From Marschall to the Ministry of Foreign Affairs, 3.2.1905, Pera, **PA/AA**, Türkei 152. See also "Französische Förderungen an die Türkei", **Frankfurter Zeitung**, 26.3.1905.
27. Ferid Pasha, **B.O.A., Y.A. - Hus.**, 483/152.
28. Cemal Paşa, **Hatıralar**, İstanbul, 1977, pp. 97, 98. From Bodman to Bülow 17.6.1904, Tarabya, **PA/AA**, Türkei 152.
29. From Marschall to the Ministry of Foreign Affairs, 7.4.1905, Pera, **PA/AA**, Türkei 152.
30. From Ranzi to the Prime Ministry and the Ministry of Foreign Affairs, 4.7.1913, Damascus, **HHStA**, Admin. Reg., F 19/33.
31. Hakkı Pasha and Ferid Pasha, **B.O.A., Y.A. - Hus.**, 473/50. Also Ferid Pasha, **Y.A. - Hus.**, 472/13. Ferid Pasha, in one of his other submissions, emphasizes that the company was subject to the Ottoman law, **Y.A. - Hus.**, 472/98.
32. Ferid Pasha, **B.O.A., Y.A. - Hus.**, 482/140.

33. **B.O.A., Y.A. - Res.**,129 /25. For similar demands, **B.O.A., Y.A. - Hus.**, 482/140.

34. "Französische Förderungen an die Türkei", **Frankfurter Zeitung**, 26.3.1905. See also Ferid Pasha, **B.O.A., Y.A. - Hus.**, 476/86. On the same issue, **Y.A. - Hus.**, 483/152.

35. For instance, Constanz had stated that while Russia lent money on 5 per cent interest, they were lending with 4 per cent, thus they expected a response for their good faith. See **B.O.A., Y.A. - Hus.**, 485/26.

36. Edmund Naumann, **Vom Goldenen Horn zu den Quellen des Euphrat,**, München und Leipzig, 1893, p. 437.

37. Hasan Pasha, First member of Komisyon-ı Ali, 12.5. 1900, **B.O.A., Y. PRK. ASK**, 161/22. For the commissioning of two engineers from the School of Engineering see Komisyon-ı Ali, 2.10.1900, **B.O.A., Y. M. MRZ.d**, 9080. For detailed information about the salaries and qualifications of these engineers see the Minister of Public Works, Zihni Pasha, 13.5. 1900, **Y. PRK. TNF**, 6/77.

38. The report by Sirkasier, 10.5.1900, **B.O.A., Y. MTV**, 202/60.

39. W. Berdrow, "Die Hedschasbahn", **ZdVDEV**, 15.12.1906, p. 1513.

40. Sait Toydemir, "Hicaz Demiryolu İnşaatı Tarihinden", **Demiryollar Dergisi**, no. 275–278, (1948), p. 65.

41. The part about Medina–Mecca and Jeddah–Mecca in Muhtar Beg's report is included in the annexes of Auler's book. See Auler, pp. 74–80. The book makes plenty of references to the report in Parts 4 and 5, pp. 29–38.

42. The aforementioned report to Governership of Syria, 23.5.1900, **B.O.A., Y. MTV**, 202/118.

43. From Lutticke to Bülow, 12.6.1902, Damascus, **PA/AA**, Türkei 152. On the other hand Auler Pasha praises Kazım Pasha's energy, organizational skills, and especially his success in resolving the conflicts between civil and military groups. See Auler, p. 26. Hejaz railway consulting engineer Kapp von Gültstein also praises the intelligence and managerial skills of Kazım Pasha. See Gültstein's report, 6.11.1905, **B.O.A., Y. PRK. TNF**, 8/35.

44. "Die geplante Mekka - Eisenbahn", **Frankfurter Zeitung**, 10.12.1900.

45. **Vossische Zeitung**, 1.3.1902.

46. According to Karal, among the reasons behind the closure of Engineering schools of *Muhendishane-i Berri-i Humayun* and *Turuku Maabir* in the Ottoman Empire were "... the fact that the majority of the students were Christian. After a while the idea of constructing a new Engineering School emerged... [At the new school] it was found appropriate for the teachers and students to wear military uniforms. Thus, with the military attire the gates of the school would be closed to the Christian children and so it would educate only Muslim engineers", Karal, pp. 398, 399.

47. İlhan Tekeli and Selim İlkin, **Osmanlı İmparatorluğu'nda Eğitim ve Bilgi Üretim Sisteminin Oluşumu ve Dönüşümü**, Ankara, 1993, p. 80.

48. Ambassador Marschall was cautious against the Meissner's efforts to bring German technical staff to the key points of the railway construction. The Ambassador's fear was that a failure resulting from the Ottomans' mismanaging of the railway would be attributed to the Germans and their "national pride" would be damaged. From Marschall to Bülow, 10.1.1903, Pera, **PA/AA**, Türkei 152.
49. For the distrust of Abdülhamid for the European countries except Germany, Sultan Abdülhamit, pp. 128, 131, 132, 137, 154, 155 et al.
50. Mygind, **Vom Bosphorus...**, p. 85.
51. From Marschall to Bülow, 25.2.1906, Pera, **PA/AA**, Türkei 152. Two weeks after this report a senior official working in the German State Railways named Schmedes was asking for permission from the German Ministry of Public Works in order to work in the Hejaz line. See from Schmedes to the Ministry of Public Works, 11.3.1906, Pera, **PA/AA**, Türkei 152.
52. Walter Pick, "Der deutsche Pionier Heinrich August Meissner Pascha und seine Eisenbahnbauten im nahen Osten 1901–1917", **Jahrbuch des Instituts für Deutsche Geschichte**, 4. Band, 1975, p. 259.
53. Wilhelm Feldmann, "Bei Meissner Pascha", **Berliner Tageblatt**, no. 31, 18.1.1917.
54. Pönicke, pp. 1, 2.
55. Pönicke, p. 9. Pick, p. 260. Also see Auler Pascha, "Besprechung", **AfEW**, Berlin, 1907, p. 316
56. Pick also has similar opinions about Meissner, Pick, p. 263. Auler, pp. 69, 70.
57. Walter Pinhas Pick, "Meissner Pasha and the construction of railways in Palestine and neighboring countries", **Ottoman Palestine 1800–1914**, ed. Gad G. Gilbar, Leiden, 1990, pp. 185, 187, et al. Pick has exaggerated Meissner's initative in this article as he did not take into account the supervision and contribution of the Ottoman bureaucracy on decision-making processes, which were apparent in the Ottoman documents.
58. Osman Erkin, "Demiryolu Tarihçesinden: Hicaz Demiryolu", **Demiryollar Dergisi**, no. 269 (1948), p. 23. After the Hejaz railway reaches Medina, Meissner Pasha will work at the Baghdad Railway for a while, Pönicke, pp. 16, 17. In World War 1 we will see Meissner undertaking again the construction of Hejaz/Egypt line in the Hejaz railway. See chapter six of this study and Feldmann's interview. Meissner who had to leave the country in 1918 had worked in Tirana for a while and then in 1924 he was invited to Turkey. Meissner who worked in the Anatolian Railways between 1927 and 1933 gave lectures at university and lived in Turkey until he passed away in 1940. See Pönicke, pp. 31, 32. Also see Erkin, p. 23. For the life and death of Meissner, Dieckmann, "Dr h.c. Meissner Pascha, der Erbauer der Hedjasbahn", **Zeitung des Vereins mitteleuropaeischer Eisenbahnverwaltungen**, 1940. Pick, on the other hand, reports the death knell in his complimentary work through these dramatic words. " Meissner Pasha died on January 14, 1940 in the capital of Turkey as a Doctor–Engineer and as a man of many titles. Five

months after his death, the first train of the Baghdad Railway, the construction of which he had directed for five years, has arrived to Baghdad from İstanbul." See Pick, p. 262.

59. Mygind, Vom **Bosporus**..., p. 85; At that time 43 engineers worked in the construction, "Die Hedschasbahn (Chemin de fer Hamidie du Hedjaz)", **ZdVDEV**, no. 89, 12.11.1904, p. 1401.

60. Ochsenwald, pp. 33, 34.

61. From Bodman to Bülow 21.8.1904, Tarabya, **PA/AA**, Türkei 152; Pick, pp. 265, 266.

62. From Minister of War to the Grand Vizierate, 5.6.1910, **B.O.A., DH. MUİ**, 101/47.

63. Charles Eude Bonin, "Le Chemin de fer du Hedjaz", **Annales de Geographic** (1909), p. 427. Also see the work of Auler Pasha that includes his second visit to the region: Auler Pascha, **Die Hedschasbahn II. Teil: Ma'ân bis El Ula**, Gotha: Justus Perthes, 1908, p. 64. As both of the Auler's works have the same title from now on we will cite the second book with its date of publication.

64. Auler, p. 25. 26 foreign engineers were working in the previous year, Guthe, p. 16

65. For instance, the decree dated July 18, 1904. **B.O.A., Y. MTV**, 262/46.

66. From the Grand Vizierate to the Ministry of Internal Affairs, 21.9.1905, **B.O.A., DH. MKT**, 1035/58.

67. Kapp von Gültstein, 6.11.1905, **B.O.A., Y. PRK. TNF**, 8/35.

68. For his memoirs that are about the Hejaz line see Mustafa Şevki Atayman, **Bir İnşaat Mühendisinin Anıları (1897–1918)**, İstanbul, 1984, pp. 46–96.

69. The answer to the Selanik Vilayet-i Celile, 20 August 1907, **B.O.A., DH. MKT**, 1158/12.

70. From the Grand Vizier to the Ministry of Internal Affairs, 19.6.1909, **B.O.A., DH. MKT**, 1306/21.

71. Tahsin Pasha, 11.7.1906, **B.O.A., İ. Hus**, 143/1324.Ca.67.

72. From Calice to Gotuchorski, 9.5.1900, İstanbul, **HHStA**, Admin. Reg., F 19/18. Komisyon-ı Ali, **B. O. A.,Y. MTV**, 204/69. On the same issue, Hasan Pasha, 12.5.1900, **B.O.A., Y. PRK. ASK**, 161/22.

73. **Vossische Zeitung**, 1.3.1902. **Die Post**, 23.1.1904. For the rail importation from Belgium, the protocol dated 13.11.1900 and numbered 5553, **B.O.A., Y. MRZ.d**, 9080.

74. "Die Hedschasbahn", **ZdVDEV**, 12.11.1904. For the efforts to produce traverses from the demesnes, Komisyon-ı Ali, 18.9.1900, **B.O.A., Y. MRZ. d**, 9080.

75. For Mehmet Beg, **B.O.A., Y. EE. KP**, 18/1794; for Mustafa, **B.O.A., Y. EE. KP**, 15/1482.

76. Pönicke, p. 6. From Schroeder to Bülow, 26.06.1905, Beirut, **PA/AA**, Türkei 152.

77. From Johannes (in the name of Stahlwerk- Verband) to Bülow, 10.6.1908, Düsseldorf, **PA/AA**, Türkei 152.

78. Pönicke, p. 6. Freight wagons were being built in Belgium, Auler, p. 42.

79. Auler (1908), p. 64. The number of locomotives in the Hejaz railway had reached up to 96 in 1913, Ruppin, p. 319.

80. Ochsenwald, p. 42.

81. From Johannes to the Ministry of Foreign Affairs, 3 and 4 March 1910. Dusseldorf, **PA/AA**, Türkei 152.

82. From a Deutsche Levante Linie official to the Ministry of Foreign Affairs, 5 March 1912, Hamburg, **PA/AA**, Türkei 152.

83. Ulrich Trumpener, "Almanya ve Osmanlı İmparatorluğu'nun Sonu", **Osmanlı İmparatorluğu'nun Sonu ve Büyük Güçler**, ed. Marian Kent, İstanbul, 1999, p. 137.

84. From Hardegg to Hollweg, 20.4.1912, Haifa, **PA/AA**, Türkei 152.

85. From Hardegg to Hollweg, 22.4.1912, Haifa, **PA/AA**, Türkei 152.

86. From Hardegg to Hollweg, 25.4.1912, Haifa, **PA/AA**, Türkei 152.

87. From Hardegg to Hollweg, 17.10.1913, Haifa, **PA/AA**, Türkei 152.

88. The Press Department Directorate, 15.6.1906, **B.O.A.DH. MKT**, 092/16. Concerning the statement to the Austrian Embassy that anyone who wanted to join the tender was free to apply, 1.11.1905, **B.O.A., Y. MTV**, 280/6.

89. Gülsoy, p. 125.

90. From Embassy to Bülow, 22.6.1904, Cairo, **PA/AA**, Türkei 152.

91. Rıfat Pasha, **B.O.A., Y.A. - Hus.**, 411/171.

92. **B.O.A., Y.A. - Hus.**, 466/136. "Die Hedschasbahn", **AfEW**, 1916, p. 290. Guthe, p. 19. H. von Kleist, "Die Hedschasbahn", **Geographische Zeitschrift**, Year 13, no. 3, (1907), p. 155. From Marschall to Bülow, 19.3.1902, Pera, **PA/AA**, Türkei 152.

93. From Keller to Schroeder, 11.2.1904, **PA/AA**, Haifa, Türkei 152.

94. **Vossische Zeitung**, 3.8.1904.

95. From Schroeder to Bülow, 26.6.1905, Beirut, **PA/AA**, Türkei 152, p. 14.

96. Ochsenwald, pp. 39, 40.

97. "Die Hedschasbahn", **AfEW**, 1916, p. 293; Auler, p. 37.

98. Kurt Becker, "Die Hedschasbahn", **Orient**, no. 5 (April 1963), p. 193.

99. Hecker, p. 1066. From Keller to Schroeder, 11.2.1904, Haifa, **PA/AA**, Türkei 152.

100. From Hardegg to Hollweg 16.7.1914, Haifa, **PA/AA**, Türkei 152, p. 20

101. **B.O.A., Y. MTV**, 262/46.

102. From Schroeder to Bülow, 26.6.1905, Beirut, **PA/AA**, Türkei 152, p. 22.

103. Z.H. Zaidi, "Chemin de fer du Hidjaz", **Encyclopedie d' Islam**, Tome II, Leiden, 1965, p. 376. Mehmed Hulusi, the head of Bridges and Roads Department, 1.3.1907, **B.O.A., Y. PRK. TNF**, 3/2.

104. From Hardegg to Hollweg, 16.7.1914, Haifa, **PA/AA**, Türkei 152, p. 27.

105. From Hardegg to Hollweg, 18.10.1911, Haifa, **PA/AA**, Türkei 152.

106. Fiedler, p. 78.

107. From Hardegg to Hollweg, 18.10.1911, Haifa, **PA/AA**, Türkei 152.

108. Fiedler, p. 79.

109. Guthe, p. 20. "Der Hafen der Mekkabahn", **Deutsche Rundschau**, Year 31, (1909), p. 330.
110. **Berliner Tageblatt**, 7.9.1911.
111. From the Austrian Embassy to Berchtold. 5.11.1913, İstanbul, **HHStA**, Admin. Reg., F 19/33.
112. From the Grand Vizier to the Ministry of Internal Affairs, 31.5.1910, **B.O.A., DH. İD**, 52/5. Also see Mehmed Hulusi, in line with the demand to have the Haifa port constructed as soon as possible, 1.3.1907, **B.O.A., Y. PRK. TNF**, 3/2.
113. **Berliner Tageblatt**, 7.9.1911.
114. The Telegraph from Haifa sent with the signs of the Acre supervisor İbrahim and his friends, **B.O.A., DH. İD**, 6/55 (abridged and abbreviated). Although the Ottoman government had decided to extend the railway from Haifa to Acre, as the residents of Acre mentioned, due to financial problems this decision had been deferred. **B.O.A., DH. İD**, 9/3.
115. For a similar comment, Alex Carmel, "Die deutsche Palaestinapolitik 1871–1914", **Jahrbuch des Instituts für Deutsche Geschichte**, 4. Band (1975), p. 254.
116. From Hardegg to Hollweg 16.7.1914, Haifa, **PA/AA**, Türkei 152.
117. From Hardegg to Hollweg 16.7.1914, Haifa, **PA/AA**, Türkei 152..
118. Auler, p. 35. For the German colony in Haifa, Saad, "Die Mekkabahn und die Stadt Hayfa am Karmelgebirge", **Petermanns Mitteilungen**, no. 51 (1905), pp. 190, 191.
119. This support continued in the coming years. For example; Max Roloff was saying "we should help them" emphasizing that the Ottomans could not be successful against the British plots on their own. Roloff also mentioned the need for the Germans to make efforts to reconcile the Turks and the Arabs, Roloff, pp. 16, 24 et al.
120. From Hardegg to Hollweg, 9.8.1913, Haifa, **PA/AA**, Türkei 152.
121. **ZdVDEV**, 9.8.1913. From Padel to Hollweg, 3.7.1911, Beirut, **PA/AA**, Türkei 152.
122. From Mutius to Wangenheim, 9.5.1914, Beirut, **PA/AA**, Botschaftsakten Konstantinopol.
123. **DLZ**, 15.11.1912.
124. From Ranzi to the Prime Ministry and the Ministry of Foreign Affairs, 4.7.1913, Damascus, **HHStA**, Admin. Reg., F 19/33.
125. From Mutius to Wangenheim, 9.5.1914, Beirut, **PA/AA**, Botschaftsakten Konstantinopol.
126. From Ranzi to Prime Ministery and Ministery of Foreign Affairs, 30.4.1914 and 5.8.1913, Damascus, **HHStA**, Admin. Reg., F 19/33.
127. About the issue being debated in Meclis-i Vukela see decree dated 11.1.1913, **B.O.A.,MV**, 173/4.
128. From Wangenheim to Hollweg, 2.6.1913, Tarabya, **PA/AA**, Türkei 152.
129. From the Consulate to Hollweg, 26.9.1913, London, **PA/AA**, Türkei 152.

130. L. Bruce Fulton, "Fransa ve Osmanlı İmparatorluğu'nun Sonu", in **Osmanlı İmparatorluğu'nun Sonu**..., p. 184.

131. From Pallavicini to the Prime Ministery, report titled as "Die Eisenbahnfragen im türkisch-französischen Übereinkommen", 22.4.1914, İstanbul, **HHStA**, Admin. Reg., F 19/33. From Pallavicini to Berchtold, 23.7.1914, Yeniköy, **HHStA**, Admin. Reg., F 19/33.

132. From Ranzi to the Ministry of Foreign Affairs, 4.6.1913, Damascus, **HHStA**, Admin. Reg., F 19/33. For the privileges the French received, see the report of Pallavicini titled "Die Eisenbahnfragen ...". See also from Pallavicini to Berchtold, 8.10. 1913, İstanbul, **HHStA**, Admin. Reg., F 19/33.

133. For the 1904 process see **B.O.A., Y.A. - Hus.**, 476/86. For 1913, from the Austrian Embassy to Berchtold, 25.10.1913, İstanbul, **HHStA**, Admin. Reg., F 19/33.

134. From Hardegg to Hollweg, 8.10.1913, Haifa, **PA/AA**, Türkei 152.

135. From Hardegg to Hollweg, 13.10.1913, Haifa, **PA/AA**, Türkei 152.

136. From Hardegg to Hollweg, 9.8.1913, Haifa, **PA/AA**, Türkei 152.

137. From Hardegg to Hollweg, 26.9.1913, Haifa, **PA/AA**, Türkei 152.

138. From Miquel to Hollweg, 29.9.1913, Cairo, **PA/AA**, Türkei 152.

139. Aforementioned report.

140. From Karl Helfferich to Foreign Ministry, 28.7.1913, Berlin, **PA/AA**, Türkei 152.

141. From Wangenheim to Hollweg, 10.10.1913, İstanbul, **PA/AA**, Türkei 152.

142. From Hardegg to Hollweg, 9.8.1913, Hayfa, **PA/AA**, Türkei 152.

143. From Hardegg to Hollweg, 21.10.1913, Hayfa, **PA/AA**, Türkei 152.

144. From Wangenheim to Hollweg, 2.6.1913, Tarabya, **PA/AA**, Türkei 152.

145. "Französische Plaene in Syrien und Palaestina", **Schlesische Zeitung**, 30.10.1913 and 29.10.1913.

146. Mandate, 8 January 1914, no. 2889, **B.O.A., MV**, 232/124. See also Layiha-ı Kanuniye signed by Mehmed Reşad, **B.O.A., İ. MLU**, 1322. N. 45.

147. From the Consulate General in Beirut to the Prime Ministry, titled "Eisenbahnen in Syrien und Palaestina", 4.5.1914, Beirut, **HHStA**, Admin. Reg., F 19/33. Also see from Hardegg to Mutius, 17.1.1914, Haifa, **PA/AA**, Botschaftsakten Konstantinopel.

148. Mandate, 4 October 1916, no. 193, **B.O.A., MV**, 245/46.

149. For the first haji convoy that was numbered more than 200 and went to Zarqa by the Hejaz railway, Ferid Pasha, **B.O.A., Y.A. -Hus.**, 464/39.

150. From the Embassy in Cairo to Bülow, 22.6.1904, Cairo, **PA/AA**, Türkei 152.

151. H. von Kleist, "Die Hedjasbahn", **Asien**, Heft 6, 1906, p. 85.

152. Komisyon-ı Ali, 28 and 31 March 1904. **B.O.A.,Y. MTV**, 258/91 and **Y. MTV**, 258/122. For a request (*ariza*) expressing a similar concern, **B.O.A., Y.EE.KP**, 27/2616.

153. Mygind, **Vom Bosporus**..., pp. 76, 77. For a similar comment, Guthe, p. 14.

154. For instance, for the telegram sent to Aydın district, **B.O.A., Y. EE. KP**, 22/2147.

256 NOTES TO PAGES 125–130

155. From Bodman to Bülow, 21.8.1904, Tarabya, **PA/AA**, Türkei 152.
156. From Marschall to Bülow, 18.9.1904, Tarabya, **PA/AA**, Türkei 152.
157. **Stamboul**, 15 and 17 August 1904.
158. From the Consulate in Beirut to Marschall, 9.9.1904, Beirut, **PA/AA**, Türkei 152.
159. "Die projektierte Eisenbahn von Djeddah nach Mekka", **Frankfurter Zeitung**, 28.10.1904.
160. **B.O.A., Y. PRK. AZJ**, 50/10.
161. From Marschall to Bülow, 18.9.1904, Tarabya, **PA/AA**, Türkei 152.
162. Eduard Mygind, **Syrien und die türkische Mekkapilgerbahn**, Halle, 1906, p. 11.
163. Koloğlu, "Muktesid Musa'ya...", p. 140.
164. **B.O.A., Y. PRK. TNF**, 8/37.
165. Auler's report titled as "Die Gründung der Teilstrecken Damaskus – Maan an der Hedschasbahn von 1 Dezember 1904", **PA/AA**, Türkei 152. For detailed information, **B.O.A., Y. MTV**, 288/40.
166. For detailed information about construction, Telegraph from Hedjaz Railway Construction Ministery, 25. 11. 1904, **B.O.A., Y. MTV**, 268/175.
167. **Hicaz Demiryolu'nun Vâridât...**, p. 1. Also see **B.O.A, MV**, 131/4.
168. Komisyon-ı Ali, 3.7.1906, **B.O.A., Y. MTV**, 288/40.
169. Meclis-i Vükela, 8.11.1908, **B.O.A., İ. TNF**, 1326. L. 2.
170. From Schroeder to Bülow, 26.6.1905, Beirut, **PA/AA**, Türkei 152, p. 2.
171. Medina Treasury Official Rashid Abdul Quadir, **B.O.A,, Y. PRK. HH**, 37/81.
172. Komisyon-ı Ali, 2.2.1907, **B.O.A., Y. MTV**, 293/96.
173. "Telegraph from the Guardianship of Medina", 26.8.1907, **B.O.A., DH. MKT**, 1194/5.
174. From an Embassy Officer to Bülow, 14 June 1908, Tarabya, **PA/AA**, Türkei 152. In the report it is stated that Marchall has sent a copy to the Emperor via the Embassy.
175. Auler (1908), pp. 47–54.
176. **B.D.** vol. 5, p. 8.
177. The Annual Report for Turkey for 1907, **B.D.** vol. 5, p. 43.
178. From Pallavicini to Aehrenthal, 6.3.1908, İstanbul, **HHStA**, Admin. Reg., F 19/18. Also see from Pallavicini to Aehrenthal, January 15, 1908, İstanbul, **HHStA**, Admin. Reg., F 19/18, p. 7.
179. The telegraph, 5.3.1907, **B.O.A., DH. MKT**, 1150/92.
180. The telegraph, 28.9.1907, **B.O.A., Y. PRK. ASK**, 251/8.
181. Meclis-i Vükela, 8.11.1908, **B.O.A., İ. TNF**, 1326. L. 2.
182. Bonin, p. 427.
183. From Padel to Bülow, 31. 8. 1908, Beirut, **PA/AA**, Türkei 152. According to a report by Komisyon-ı Ali, 20.5.1908, a high number of notables coming from Egypt must have attended the ceremonies, **B.O.A., Y. MTV**, 309/156.
184. "Izzat Pasha and The Hedjaz Railway", The Times, 3.9.1908.

185. From Pallavicini to Aehrenthal, 27.7.1908, Yeniköy, **HHStA**, Admin. Reg., F 19/18. The support of the Muslim world to the Hejaz railway will be interrupted by the dethronement of Abdülhamid in 1909 and later when all the doubts about the approach of the members of *İttihat ve Terakki* (Union and Progress) towards Islam and caliphate disappeared the relations would turn to normal. For details see Özcan, pp. 188, 197, 198 et al.

186. Ochsenwald, p. 77.

187. The **Times of India**, 18 August 1908.

188. From The General Consulate in Calcutta to Bülow, 3 September 1908, Simla, **PA/AA**, Türkei 152.

189. From Schroeder to Bülow, 26.6 1905, Beirut, **PA/AA**, Türkei 152, p. 10.

190. Auler, p. 44. Also see "Die Hedschazbahn", **Gaea Natur und Leben**, no.10, (1907), p. 620.

191. From Jenisch to Bülow, 3.6.1905, Cairo, **PA/AA**, Türkei 152.

192. **B.O.A., Y. PRK. TNF**, 8/35.

193. Auler, p. 44. Auler (1908), p. 58. Schlagintweit, "Die Mekkabahn", p. 129. For similar efforts, from Schroeder to Bülow, 26.6.1905, Beirut, **PA/AA**, Türkei 152, pp. 11–13.

194. Imbert, p. 84. Pönicke, p. 8. Auler (1908), p. 57.

195. Komisyon-ı Ali., **B.O.A., Y. MTV**, 262/46.

196. Meclis-i Vükela, 18.9.1907, **B.O.A., MV**, 117/13.

197. Faroqhi, Hacılar..., pp. 53, 54.

198. Komisyon-ı Ali, 1.11.1905, **B.O.A., Y. MTV**, 280/6.

199. Faroqhi, Hacılar..., pp. 51, 52.

200. Auler (1908), p. 57.

201. **B.O.A., Y. PRK. KOM**, 15/5.

202. Atayman, pp. 55–8.

203. İsmail Hakkı Uzunçarşılı, **Mekke-i Mükerreme Emirleri**, Ankara, 1972, p. 57.

204. Atayman, p. 47. Auler, pp. 33, 34. For the amount of rain per m^2 in the region in 1913–14, P. Dieckmann, "Ergebnisse der Regenmessung im Hedschaz-bahngebiet, Winter 1913/14", **Zeitschrift des Deutschen Palaestinaver-eins**, Band 37 (1914).

205. From Jenisch to Bülow, 3.6. 1905, Cairo, **PA/AA**, Türkei 152.

206. **B.O.A., Y.PRK.BŞK**, 62/60. Auler, pp. 46, 47. Hecker, p. 1312.

207. **B.O.A., Y. MTV**, 285/169.

208. **B.O.A., Y. MTV**, 290/128.

209. Auler, p. 46.

210. Blanckenhorn, "Die Hedschaz – Bahn, auf Grund ...", p. 227.

211. Kapp von Gültstein, 6.11.1905, **B.O.A., Y. PRK. TNF**, 8/35.

212. Seraskier Rıza Pasha, 3.6.1906, **B.O.A., Y. MTV**, 287/84.

213. From the Grand Vizier to Ministry of Internal Affairs, 12.3.1906, **B.O.A., Y. MTV**, 451/1.

214. Auler, p. 26.

215. From Jenisch to Bülow, 3.6.1905, **PA/AA**, Türkei 152.
216. Auler, p. 50.
217. **B.O.A., Y. MTV**, 226/25.
218. Auler (1908), pp. 58, 59.
219. Seraskier Rıza Pasha, 29.5.1906, **B.O.A., Y. MTV**, 287/34.
220. **B.O.A., Y. MTV**, 292/21. **B.O.A., Y.PRK. TKM**, 50/59. Also see Seraskier Rıza Pasha, 13.2 1907, **B.O.A., Y. MTV**, 294/134.
221. Auler, pp. 48,49.
222. Auler (1908), p. 28.
223. Paul Rohrbach, "Bagdadbahn und Mekkabahn", Übersee, Nr. 4, 26.1.1908, p. 2.
224. Kapp von Gültstein, 6.11.1905, **B.O.A., Y. PRK. TNF**, 8/35.
225. Metin Hülagü, **Bir Umudun İnşası, Hicaz Demiryolu**, İstanbul, 2008, p. 94. Rohrbach, "Bagdadbahn... ", p. 2.
226. Auler, p. 51.
227. Ochsenwald, p. 37.
228. Auler, p. 51; Auler (1908), p. 61.
229. **B.O.A., Y. MTV**, 226/25.
230. Regulation of healthcare services available for Komisyon-ı Ali health officers and other workers, **B.O.A., İ. Hus**,1322.L.49.
231. Gülsoy, pp. 121, 122.

Chapter 5   Forces Resisting the Hejaz Railway

1. This issue was mentioned in chapter two of this book. See pp. 60–68.
2. Auler, p. 47. The importance of the matter was also mentioned by Gültstein in 1901, **B.O.A., Y. PRK. TNF**, 8/35.
3. "Die neue Mekkabahn", **Die Katholischen Missionen**, Nr. 10 (1906/07), p. 219.
4. From Oppenheim to Bülow, 8.4.1905, Cairo, **PA/AA**, Türkei 152.
5. **B.O.A., Y.A. - Hus.**, 482/160. Egyptian Ambassador of the time also reports Pasha'a opinions on this subject. See report from Jenisch to Bülow, 28.12.1904, Cairo, **PA/AA**, Türkei 152.
6. Fiedler, pp. 85, 86. See also Dieckmann, "Die Verkehrsgeographische Bedeutung der Vierlaenderecke bei Akaba", AfEW (1943), p. 135.
7. Ferid Paşa, **B.O.A., Y.A. - Res.**, 129/3.
8. From Schroeder to Bülow, 26.6.1905, Beirut, **PA/AA**, Türkei 152.
9. From Schroeder to Bülow, 7.4.1905, Beirut, **PA/AA**, Türkei 152.
10. From Schroeder to Bülow, 21.2.1906, Beirut, **PA/AA**, Türkei 165.
11. Kleist, "Die Hedschas-Bahn", **Geographische Zeitschrift**, 1907, pp. 155, 156; "Orientalia", **Die Zukunft**, Nr. 47, 22.8.1908, p. 283. See also **B.O.A., Y. PRK. SGE**, 10/61.

12. Seraskierate, 27.1.1906, **B.O.A., Y. PRK. ASK**, 236/38.
13. "The Sinai Boundary Dispute and the British Naval Demonstration, 1906", **B. D.**, vol. V, p. 190.
14. Report titled "Problem of Khedive and Aqaba" from Oppenheim to Bülow, 30.3.1906, Cairo, **PA/AA**, Türkei, 165.
15. **B.O.A., Y. PRK. SGE**, 10/61.
16. Hermann's translation of the article in *Morning Post*, **B.O.A., Y. PRK. TKM**, 49/44.
17. Mustafa Kâmil's article translated by Hermann, **B.O.A., Y.PRK. TKM**, 49/44.
18. Tahsin Pasha, p. 28.
19. **B.D.**, vol. V, p. 189.
20. Graves, p. 79.
21. **Berliner Tageblatt**, 6.3.1906.
22. **B.D.**, p. 190.
23. **B.O.A., Y. PRK. TKM**, 49/44. Rohrbach described this situation as the most prominent reaction shown by the British against Hejaz railway, see "Bagdadbahn und...", p. 3.
24. Graves, p. 83.
25. Memorandum by Mr Fitzmaurice, Constantinople, August 10, 1913, **B.D.**, vol. X, Part I, p. 512.
26. "Die eisenbahnlose Zone der Englaender", **Neue Preusische Zeitung**, 8.4.1908.
27. From Metternich to Bülow, 23.4.1906, London, **PA/AA**, Türkei 152.
28. From Marschall to Bülow, 22.4.1906, Pera, **PA/AA**, Türkei 165. The memoir of an Ottoman administrator who visited Kaiser Wilhelm is also worth mentioning: "German Emperor addressed to Prime Minister Bülow: 'Do you know that since Hejaz railway is so close to the sea, the necessity to build a branch line to the port is now obvious? Due to such necessity, a branch line from Maan to Aqaba was planned to be built. Upon hearing this, the British obstructed'. In return, the Prime Minister and Emperor smiled ironically." Mehmet Raci, member of Military Commission, 5 April 1906, **B.O.A., Y. PRK. KOM**, 15/5.
29. Ernst Schick, "Die Hedschasbahn", **Technisches Magazin**, no. 6 (1910), p. 453.
30. Ferid Paşa, **B.O.A., Y.A. - Hus.**, 471/41.
31. Paul Rohrbach, **Die Bagdadbahn**, Berlin 1911, pp. 18, 19.
32. From Marschall to Hollweg, 30 May 1910, **G.D.D.**, vol. III, pp. 390, 391.
33. Trumpener, "Almanya...", p. 140.
34. Oppenheim's report, 12.1.1907, Cairo, **PA/AA**, Türkei 152.
35. Oppenheim's report, 22.6.1907, Cairo, **PA/AA**, Türkei 152. Oppenheim's deep interest in the matter together with his frequent visits to Maan, made the British feel uneasy about the matter. See Haslip, **II. Abdülhamid**, pp. 271, 272.

36. From Lascelles to Grey, 17 May 1906, Berlin, **B.D.**, vol. III, p. 340.
37. **B.O.A., Y.A. - Hus.**, 482/160. From Jenisch to Bülow, 28.12.1904, Cairo, **PA/AA**, Türkei 152.
38. Tahsin Pasha, p. 28.
39. Mim Kemal Öke, **II. Abdülhamid ve Dönemi**, İstanbul, 1983, p. 129.
40. From Schroeder to Bülow, 20.4.1908, Beirut, **PA/AA**, Türkei 152.
41. Grand Vizier Ferid Pasha, **B.O.A., Y.A. - Hus.**, 420/22.
42. **B.O.A., DH. İD**, 134/3.
43. From Padel to Foreign Affairs, 28.5.1908, Pera, **PA/AA**, Türkei 152. For work on building a quarantine centre in Medine, **B.O.A., Y. PRK. BŞK**, 226/109.
44. From the Undersecretary to Internal Affairs, 25 January 1914, **B.O.A., DH. İD**, 191/9.
45. Urgent telegram sent from Hejaz railway General manager to Internal Affairs, 24.1.1911, **B.O.A., DH.İD**, 7–1/29. Cevad Bey also warned the authorities about the possible cram in front of Tabuk Quarantine. **B.O.A., DH. İD**, 7–2/25. The British believed Tabuk as not a good choice for the main quarantine center, "Annual Report on Turkey for the year 1910", **B.D.F.A.**, vol. 20, p. 200.
46. To Ministry of Military Education and Ministry of Mail and Telegram, 3 February 1907, **B.O.A., DH. MKT**, 1140/88. The commission's decision to open quarantine centres besides Tabuk, 20.5.1913, **B.O.A., İ. MMS**, 1331.C.31. For British comment on the subject, **B.D.F.A.**, vol. 20, pp. 156, 157, 200, 201.
47. Auler, pp. 59, 77, 79, 80. In 1908 British annual reports on Ottomans, this area was referred to as "more suitable, safer, and more abundant in terms of water sources". See **B.D.**, vol. V, p. 283.
48. From Hardegg to Hollweg, 22.4.1912, Haifa, **PA/AA**, Türkei 152.
49. For the translation of Gültstein's report, 6 November 1905, **B.O.A., Y. PRK. TNF.**, 8/35.
50. Aforementioned report. Auler also makes reference to the same report. Auler, pp. 59–61.
51. "Die projektierte Eisenbahn von Djeddah nach Mekka", **Frankfurter Zeitung**, 28.10.1904.
52. In the report by Mecca Amirate dated 4.2.1911, it wrote, "There is no doubt that 80 km-long Jeddah-Mecca Line may be built with no difficulty in 4–5 months". **B.O.A., DH. İD**, 9/7.
53. Similar views were shared in the reports sent by German consulate from the region. For instance, from Hardegg to Hollweg, 22.4.1912, Haifa, **PA/AA**, Türkei 152.
54. From Hejaz Railway General Manager Agent Eşref to the Grand Vizier, 20 November 1912, **B.O.A., DH. İD**, 4–2/25.
55. For annual payments (*surra*) sent to the local people in 1904, **B.O.A., DH. MKT**, 897/75.

56. "Die Hedjaz-Eisenbahn", **Mitteilungen der k. k. geographischen Gesellschaft in Wien**, XLVIII, (1904), p. 52. For the payments made to Bani Harb tribe by both Ottoman and Egyptian governments, B. Moritz, "Der Hedjâz und die Strasse von Mekka nach Medina", **Zeitschrift der Gesellschaft für Erdkunde zu Berlin**, vol. 25 (1890), p. 149.

57. Philipp, p. 68 et al.

58. From Richthofen to Bülow, 1.7.1908, Simla, **PA/AA**, Türkei 152. Richthofen criticizes such applications for the short run.

59. From Schroeder to Bülow, 26.6.1905, Beirut, **PA/AA**, Türkei 152, pp. 19, 20.

60. To the Ministry of Mail and Telegram, 24 May 1902, **B.O.A., DH. MKT**, 463/23.

61. From Lütticke to Bülow, 20.3.1901, Damascus, **Pol Archiv**, Türkei 152. For a similar news article see **Berliner Tageblatt**, 22.2.1901.

62. Commander of Medina Ferik (General) Bahri Pasha, 9 March 1909, **B.O.A., DH. MUİ**, 1–4/2.

63. From Johannes (on behalf of Stahlwerks - Verband) to Bülow, 10.6.1908, Düsseldorf, **PA/AA**, Türkei 152.

64. Meclis-i Vükela, 27 October 1907, **B.O.A., İ. AS.**, 1325. N. 15. (abridged).

65. Bonin, pp. 428, 429.

66. To General Secretary and Prime Ministry, 8 February 1908, **B.O.A., DH. MKT**, 1226/70. Philipp, pp. 66, 67. Ochsenwald, pp. 124, 125.

67. Komisyon-ı Ali, 11 February 1908, **B.O.A., Y. MTV**, 306/34. See also Kazım Pasha, **B.O.A., Y. MTV**, 305/78.

68. From Schroeder to Bülow 1.6.1908, Beirut, **PA/AA**, Türkei 152. For detailed information on the assaults see **B.O.A., DH. MKT**, 1226/70.

69. From Padel to Bülow, 19.7.1908 and 31.8.1908, Beirut, **PA/AA**, Türkei 152. Generalkommando, 16.4.1908, **B.O.A., Y. MTV**, 308/97. For new troops planned to be sent, see Komisyon-ı Ali, 23.8.1908, **B.O.A., Y. PRK. ASK**, 258/36. For soldiers sent from Trebizond, see coded telegram from Trebizond, 17.5.1908, **B.O.A., Y. PRK. UM**, 81/48.

70. Musil, **Im nördlichen Hegaz**, pp. 19, 20.

71. From Schroeder to Bülow, 1.6.1908, Beirut, **PA/AA**, Türkei 152. For armed cars, **B.O.A., DH. İD**, 1/1. For the machine gun imported from France, **B.O.A., Y. PRK. UM**, 81/48.

72. Bahri Pasha, 9 March 1909, **B.O.A., DH. MUİ**, 1–4/2.

73. Kazım Pasha's coded telegram, 2 January 1908, **B.O.A., Y. PRK. ASK**, 251/8.

74. From Bahri Pasha to Internal Affairs, 31.12.1908. **B.O.A., Y. DH. MUİ**, 1-1/38.

75. From Rashid Bin Nasır to Prime Ministry, 25 Mart 1909, **B.O.A., Y. DH. MUİ**, 1–3/25.

76. Mecca Amirate, 4.2.1911, **B.O.A., DH. İD**, 9/7.

77. Coded telegram from Commander of Medina to Internal affairs, 31.12.1908, **B.O.A., Y. DH. MUİ**, 1–1/38. Gülsoy, p. 136.

78. Musil, **Im nördlichen**..., p. 21.
79. Coded telegram by Bahri Pasha, 3 January 1908, **B.O.A., Y. DH. MUİ**, 1-1/38.
80. A.J.B. Wavell, F.R.G.S., **A Modern Pilgrim in Mecca and a Siege in Sanaa**, London, 1912, pp. 58–63.
81. Orhan Koloğlu, **Bedevi**..., p. 49.
82. Tahsin Pasha, pp. 349, 350.
83. From Schroeder to Bülow, 1.6.1908, Beirut, **PA/AA**, Türkei 152.
84. From Graf Mensdorf to Aehrenthal, London, **HHStA**, Admin. Reg., F 19/33.
85. From Hardegg to Hollweg, 19.7.1913, Haifa, **PA/AA**, Türkei 152.
86. Soleman Pasha, 17.2.1909, **B.O.A., DH. MUİ**, 1–1/37.
87. Bahri Pasha, 9 March 1909, **B.O.A., DH. MUİ**, 1–4/2.
88. Philipp, p. 68.
89. Austrian Charge d'affairs Giesl, "Eisenbahn- und Kommunikationsbauten, dann Kulturarbeiten in der Türkei mit Abschluss des Jahres 1909", 23.11.1909, **HHStA**, Admin. Reg., F 19/32, p. 33.
90. First Subsidiary of Constant Maintenance Engineering, 4.5.1910, **B.O.A., DH. MUİ**, 101/1.
91. From Schroeder to Bülow, 1.6.1908, Beirut, **PA/AA**, Türkei 152.
92. Alois Musil, **The Northern Hegaz**, New York, 1926, pp. 9, 10.
93. For an example, Bahri Pasha, 21.12.1908, **B.O.A.,DH. MUİ**, 1–1/38.
94. Gülsoy, p.141. Head of *Meclis-i Ayan* (Senate) Said Pasha, 16 March 1910, **B.O.A., İ. MLU**, 1328.Ra.5. From Ranzi to Prime Ministry and Foreign Affairs, 4.10.1912, Damascus, **HHStA**, Admin. Reg., F 19/33.
95. Faroqhi, **Hacılar**..., pp. 82, 83.
96. Uzunçarşılı, p. 23 et al.
97. "Die projektierte Eisenbahn von Djeddah nach Mekka", **Frankfurter Zeitung**, 28.10.1904.
98. Max Freiherr von Oppenheim, unter Mitarbeitung von E. Braunlich und W. Caskel, **Die Beduinen, Band II: Die Beduinenstamme in Palaestina, Transjordanien, Sınai, Hedjaz**, Leipzig, 1943, p. 322.
99. Uçarol, pp. 244, 245.
100. İ. Hakkı Uzunçarşılı, "Hicaz Vali ve Kumandanı Osman Nuri Paşa'nın Uydurma Bir İrade İle Mekke Emiri Şerif Abdülmuttalib'i Azletmesi" (issue no. 39 published apart from **Belleten**), Ankara, 1946, pp. 503–5, 512, et al.
101. Sırma, "Ondokuzuncu Yüzyıl...", p. 186.
102. Gülsoy, p. 216.
103. Uzunçarşılı, **Mekke-i Mükerreme**..., p. 28.
104. From Schroeder to Bülow, 1.6.1908, Beirut, **PA/AA**, Türkei 152.
105. Instruction from Yaver-i Ekrem Arif Pasha, 2 March 1908, **B.O.A., Y. EE**, 13/28.
106. "A summary of an important letter from Mecca", 4 April 1908, **B.O.A., Y. EE**, 13/29. The amir and governor, with the excuse that Bedouins would

rebel, had argued that the Medina–Mecca line should not be built, **B.O.A.**, **Y.PRK. UM**, 80/88 and 81/15. See also Gülsoy, pp. 214–7.

107. Komisyon-ı Ali, **B.O.A.**, **Y. MTV**, 309/30.
108. Coded telegram sent to Hejaz Province on 9 May 1908, **B.O.A.**, **Y. MTV**, 309/30.
109. Kayalı, p. 68.
110. From Padel to Bülow, 21.9.1908, Beirut, **PA/AA**, Türkei 152. See also from Toncic to Aehrentahl, 30.3.1909, Jeddah, **HHStA**, Admin. Reg. F 19/18.
111. **B.D.**, vol. V, p. 300.
112. From Toncic to Foreign Affairs, 30.3.1909, Jeddah, **HHStA**, Admin. Reg. F 19/18.
113. Uçarol, p. 244.
114. Dawn, p. 4.
115. David Fromkin, **A Peace To End All Peace**, New York, 1989, p. 112.
116. **Memoirs of King Abdullah of Transjordan**, ed. Philip P. Graves, London, 1950, pp. 61, 62.
117. Dawn, pp. 9, 13.
118. **King Abdullah**, p. 89.
119. From Mailet to Edward Grey, 13.3.1914 (enclosure), Constantinople. **B.D.**, vol. 10/2, p. 829.
120. From Pallavicini to Prime Ministry, 24.5.1910, İstanbul, **HHStA**, Admin. Reg., F 19/33.
121. From Toncic to Aehrenthal, 23.12.1910, Jeddah, **HHStA**, Admin. Reg., F 19/33.
122. Kayalı, pp. 176, 177.
123. From Strempel to Ministry of War, 7.5.1912, İstanbul, **PA/AA**, Türkei 152.
124. From Toncic to Berchtold, 14.3.1912, Jiddah, **HHStA**, Admin. Reg., F 19/33. Governor of Beirut also mentioned the idea that any intrigues displayed by amir of Mecca should be taken into consideration, from Hardegg to Hollweg, 22.4.1912 and 29.5.1913, Haifa, **PA/AA**, Türkei 152.
125. Mahmut Şevket Paşa, **Mahmut Şevket Paşa'nın Günlüğü**, İstanbul, 1988, p. 19.
126. Gülsoy, p. 221.
127. Kurşun, pp. 598–605.
128. Shakib Arslan's letter, 30 November 1913, **B.O.A.**, **DH. KMS**, 63/53.
129. Arslan's letter, 4 December 1913, **B.O.A.**, **DH. KMS**, 63/53.
130. Arslan's letter, 29 December 1913, **B.O.A.**, **DH. KMS**, 63/53.
131. Arslan's letter, 28 February 1914, **B.O.A.**, **DH. KMS**, 63/53.
132. The law on province rule ordered a governor appointed to a province from İstanbul take the chairmanship in the assembly gathered locally. With this law, Ittihadists' aim was to both satisfy the provinces' desire to act autonomously and not to destroy the unity of the state at the same time. See Feroz Ahmad, **İttihat ve Terakki (1908–1914)**, İstanbul, 1984, pp. 224, 225 et al.

133. Yusuf Hikmet Bayur, **Türk İnkılâbı Tarihi**, vol. III, Part III, Ankara, 1991, pp. 236, 237.
134. Feridun Kandemir, **Medine Müdafaası**, İstanbul, 1991, p. 423.
135. **King Abdullah**, pp. 108, 109.
136. Dawn p. 17. As it can once more be seen, we need to make reference to how loans taken from France played a restrictive role on any political decisions to be taken by the Ottoman state.
137. Fahreddin Paşa, "Şerif Hüseyin,Vehib Paşa ve Yemen" (in Kandemir, **Medine Müdafaası**). For the part summarized here, pp. 423–9; for Fahreddin Pasha's writings, pp. 421–70.
138. Dawn, p. 17.
139. **Memoirs of King Abdullah**, pp. 115, 116.
140. **King Abdullah**, pp. 120, 121.
141. Kayalı, pp. 207, 208.
142. İsmail Arar, "Macera Dolu Bir Hayat", **Tarih ve Toplum**, no. 48 (1987), pp. 361, 362.
143. Philip H. Stoddard, **Teşkilât-ı Mahsusa**, İstanbul, 1993, p. 117.
144. From Hardegg to Hollweg, Haifa, 16.4.1912, **PA/AA**, Türkei 152.
145. Beirut Provincial General Assembly, 22 December 1910, **B.O.A., DH. İD**, 52/13.
146. Komisyon-ı Ali, 12 March 1907, **B.O.A., Y. MTV**, 295/143.
147. **DLZ**, 1.5.1914.
148. In an article on the advantages of this line, Dieckmann mentioned the fact that 25,000 tourists visit the area every year. See "Die Zweiglinie Affule-Jerusalem der Hedschazbahn", **Zeitschrift des Deutschen Palaestinavereins**, vol. 37, (1914), pp. 269, 270.
149. From Hardegg to Hollweg, 16.4.1912, Haifa, **PA/AA**, Türkei 152.
150. Aforementioned report.
151. From Padel to Bülow, 26.9.1908, Beirut, **PA/AA**, Türkei 152. For a report depicting all the advantages of Jaffa–Jerusalem railway we have mentioned so far, see the report from Wangenheim to Hollweg, 2.6.1913, Tarabya, **PA/AA**, Türkei 152.
152. From General Staff to Internal Affairs, 21.11.1911, **B.O.A., DH. İD**, 4–1/38.
153. **B.O.A., DH İD**, 4–2/17.
154. The application made by an officer from Deutsche Levante Linie to Foreign Affairs, 5.3.1912, Hamburg, **PA/AA**, Türkei 152. In Chapter 4, government/capital relations were held in detail in terms of German investments on Ottoman lands.
155. "Eröffnung einer Zweiglinie der Hedschasbahn", **AfEW**, 1913, p. 815. See also **DLZ**, 1.3.1913.
156. From Hardegg to Hollweg, 2.5.1914, Haifa, **PA/AA**. Botschaftsakten Konstantinopel.

157. From Ranzi to Prime Ministry and Foreign Affairs, 19.12.1913, Damascus, HHStA, Admin. Reg., F 19/33.
158. From Beirut General Consulate to Prime Ministry and Foreign Affairs, 22.5.1914, Beirut, HHStA, Admin. Reg., F 19/33.
159. Cemal Pasha, Hatıralar, İstanbul, 1977, pp. 97, 98.
160. Özyüksel, Anadolu..., p. 136.
161. Ferid Pasha, 10 October 1904, B.O.A., Y.A. - Hus., 479/62. See also "Die Hamidie-Eisenbahn im Hedschas", Illustrierte Zeitung, Nr. 3168, 17.4.1904.
162. Mahmut Şevket, pp. 186, 187.
163. From Strempel to the Ministry of War, 16.5.1907, İstanbul, PA/AA, Türkei 152.
164. Martin Hartmann, "Die Mekkabahn", Orientalische Literatur-Zeitung, no.1, 15.1.1908, p. 4. Kleist, "Die Hedjasbahn", Globus, Nr. 11, 22.4.1906, p. 170.
165. From Toncic to Aehrenthal, 2.2.1911, Jeddah, HHStA, Admin. Reg., F 19/33.
166. Robert Deutsch, "Bericht über die Yemenbahn", August 1912, 56372/8 b, Wien, HHStA, Admin. Reg., F 19/33.
167. German Roloff's supportive attitude towards France and their control over railways to harm the power of the British in Yemen should be evaluated in this respect, Roloff, Arabien..., pp. 20, 21.
168. Deutsch's report.
169. Mahmut Şevket, pp. 49, 83, 85.
170. For example, see Roloff, p. 18.
171. "Die Vollendung der Hedschasbahn", AfEW, 1912, p. 1342.
172. From Hardegg to Hollweg, 16.7.1914 and 26.9.1913, Haifa, PA/AA, Türkei 152.
173. From Hardegg to Hollweg,, 17.10.1913 and 18.10.1911, Haifa, PA/AA, Türkei 152.
174. From Consulate to Aehrenthal, 16.6. 1911, Beirut, HHStA, Admin. Reg., F 19/33.
175. P. Dieckmann, "Die Bautaetigkeit im Gebiete der Hedschasbahn", ZdVDEV, 25.9.1912, p. 1179; See also "Die Hedschasbahn", DLZ, Nr. 6, 1.5.1912, p. 12.
176. From Ranzi to Prime Ministry, 4.9.1912, Damascus, HHStA, Admin. Reg., F 19/33.
177. From Ranzi to Prime Ministry, 4.6.1913, Damascus, HHStA, Admin. Reg., F 19/33.

Chapter 6   Were the Expectations Fulfilled?

1. Germans were followed by Austrians and the French, Auler, p. 39.
2. Ochsenwald, p. 96.

3. Neue Zürcher Zeitung, 21.2.1910.
4. From Padel to Bülow 21.9.1908, Beirut, PA/AA, Türkei 152. Martin Hartmann, "Die Mekkabahn,...", p. 150; Pick, p. 260; Pönicke, p. 15.
5. From Embassy to Aehrenthal, 15.7.1910, Yeniköy, HHStA, Admin. Reg., F 19/33.
6. From Cevad Bey to Internal Affairs, 29 March 1910, B.O.A., DH. İD, 52/11.
7. From Hardegg to Hollweg, 22.3.1912, Haifa, PA/AA, Türkei, 152.
8. From Pallavicini to Aehrenthal 26.1.1910, İstanbul, HHStA, Admin. Reg., F 19/33.
9. From Hardegg to Hollweg, 22.3.1912, Haifa, PA/AA, Türkei 152.
10. From Cevad Beg to Internal Affairs, 29 March 1910, B.O.A., DH. İD, 52/11.
11. Musil, Im nördlichen..., pp. 4, 5.
12. "Die Hedschasbahn", AfEW, 1916, p. 295.
13. From Hardegg to Hollweg 16.7.1914, Haifa, PA/AA,Türkei 152, p. 20. Paschasius, "Die Hedschasbahn", AfEW, 1927, p. 1147.
14. Soldiers traveled the same distance in 22–25 km/h, "Die türkischen Eisenbahnen im Kriege", ZdVDEV, Nr. 12, (11.2.1914), s. 195.
15. Hecker, p. 1550.
16. Adham M. Fahm, "Between Mystical and Military: The Architecture of the Hejaz railway (1900–1908)", Essays in Honour of Aptullah Kuran, ed. Ç. Kafescioğlu- L. Thys- Şenocak, İstanbul, 1999.
17. Gülsoy, pp. 142, 143.
18. P. Levy, "Die Betriebsmittel der Hedjazbahn", Organ für die Fortschritte des Eisenbahnwesens in technischer Beziehung, no. 6 (15.3.1911), p. 100.
19. Fiedler, pp. 112, 113.
20. If the same calculation is made for 1911, the same results are derived. See from Ranzi to Prime Ministry and Foreign Affairs, 4.10.1912, Damascus, HHStA, Admin. Reg., F 19/33.
21. From Hardegg to Hollweg, 16.7.1914, Haifa, PA/AA, Türkei 152, pp. 20, 22. In another report, Hardegg claimed that the line would be more profitable if no management errors were made. See from Hardegg to Hollweg, 22.3.191, Haifa, PA/AA, Türkei, 152. For a detailed list of expenses, Hardegg's report, 20.4.1912.
22. From Ranzi to Prime Ministry and Foreign Affairs, 30.4.1914, Damascus, HHStA, Admin. Reg., F 19/33.
23. From Hardegg to Hollweg 16.7.1914, Haifa, PA/AA, Türkei 152, p. 33.
24. Aforementioned report., pp. 37, 38.
25. From Ranzi to Prime Ministry and Foreign Affairs, 4.10.1912, Damascus, HHStA, Admin. Reg., F 19/33.
26. From Hardegg to Hollweg 16.7.1914, Haifa, PA/AA, Türkei 152, pp. 38–40.
27. To Health Minister on the quarantine centres and health control measures, 1 November 1911, B.O.A., DH. İD, 134/3.

28. From Hardegg to Hollweg 16.7.1914, Haifa, **PA/AA**, Türkei 152, pp. 29, 32, 33. As for the numbers issued by the British for 1909–1910 season, **B.D.F.A.**, vol. 20, pp. 156, 200.

29. This average was based on data from the report: From Hardegg to Hollweg 16.7.1914, Haifa, **PA/AA**, Türkei 152, p. 29.

30. From Toncic to Berchtold, 14.3.1912, Jeddah, **HHStA**, Admin. Reg., F 19/33.

31. Some radical Muslims in this group believed that easy transport to Mecca would diminish the worth of pilgrimage, and this belief led to an opposition against Hejaz railway or any other easy way of travelling. See from Schroeder to Bülow, 15.3.1904, Beirut, **PA/AA**, Türkei 165.

32. From Hardegg to Hollweg 16.7.1914, Haifa, **PA/AA**, Türkei 152, pp. 29, 32.

33. For instance, half of 180,000 pilgrims in 1910 preferred to go there over Jeddah. See from Mertens to Hollweg, 11.1.1912, İstanbul, **PA/AA**, Türkei 152.

34. From Hardegg to Hollweg 16.7.1914, Haifa, **PA/AA**, Türkei 152. For those who go on pilgrimage by Hejaz railway, see p. 29; for those who return, p. 32.

35. From Toncic to Berchtold, 14.3.1912, Jeddah, **HHStA**, Admin. Reg., F 19/33.

36. **Civil and Military Gazette**, 14.9.1904.

37. Arif Pasha, 2 March 1908, **B.O.A., Y. EE**, no. 13/28.

38. From Hardegg to Hollweg, 16.7.1914, Haifa, **PA/AA**, Türkei 152, pp. 30, 32.

39. From Schroeder to Bülow, 20.4.1908, Beirut, **PA/AA**, Türkei, 152.

40. From Cevad Bey to Internal Affairs, 5 February 1910, **B.O.A., DH. İD**, 52/11.

41. From Hardegg to Hollweg 16.7.1914, Haifa, **PA/AA**, Türkei 152, p. 34. See also, "Die Hedschasbahn", **AfEW**, 1916, p. 3111.

42. From Governor of Syria, 1 March 1910, **B.O.A., DH. İD**, 52/11.

43. From Cevad Beg to Internal Affairs, 26 March 1910, **B.O.A., DH. İD**, 52/11.

44. From Cevad Beg to Internal Affairs, 20 March 1910 and 26 March 1910, **B.O.A., DH. İD**, 52/11.

45. From Cevad Beg to Internal Affairs, 27 April 1910, **B.O.A., DH. İD**, 52/11.

46. From Basri Pasha to Internal Affairs, 9 March 1911, **B.O.A., DH. İD**, 52/11.

47. From Basri Pasha to Internal Affairs, 29 February 1912, **B.O.A., DH. İD**, 52/11.

48. From Hardegg to Hollweg, 16.7.1914, Haifa, **PA/AA**, Türkei 152, p. 30.

49. "Die Hedschasbahn", **AfEW**, 1916, p. 313; Dieckmann, "Die Pilgerbeförderung auf der Hedschasbahn", **ZdVDEV**, 30.4.1913, p. 546.

50. From Pallavicini to Aehrenthal, 6.3.1908, İstanbul, **HHStA**, Admin. Reg., F 19/18, pp. 32, 33.

51. Philipp, p. 57.

52. From Schroeder to Bülow, 26.6.1905, Beirut, **PA/AA**, Türkei 152, p. 15. For a similar analysis, **B.O.A., Y. PRK. TKM**, 50/59. Report by Algeria

Governor Abidin Pasha on how Ibn Saud provoked soldiers and how they were obliged to be sent via Hejaz railway, **B.O.A., Y. PRK. UM**, 75/87.

53. Traveler Euting reports that the village was adandoned in 1884, Auler (1908), p. 11.
54. Ochsenwald, pp. 136, 137.
55. "Annual Report on Turkey for the year 1910", **B.D.F.A.**, vol. 20, p. 200. See also Gülsoy, p. 237.
56. From Marschall to Hollweg, 9.12.1910, Pera, **PA/AA**, Türkei 177. Ochsenwald, pp. 127, 128. See also **B.D.F.A.**, vol. 20, p. 200.
57. From Hardegg to Hollweg 16.7.1914, Haifa, **PA/AA**, Türkei 152. p. 23.
58. From the Command of Medina to Internal Affairs, 8 March 1914, **B.O.A., DH. İD**, 65/47.
59. Gülsoy, pp. 239, 240.
60. From Pallavicini to Prime Ministry and Foreign Affairs, 3. 8.1915, İstanbul, **HHStA**, Admin. Reg., F 19/33.
61. Report dated 4 October 1916, no. 193, **B.O.A., MV**, 245/46.
62. Atayman, p. 52.
63. Dieckmann, "Die Hedschasbahn im Weltkriege und ihre gegenwaertige Lage", **ZdVDVE**, Nr. 48, 1.12.1921, pp. 894, 895.
64. From Padel to Wangenheim, 1.2.1915, Damascus, **PA/AA**, Türkei 152.
65. From Schmidt to Wangenheim, 8.2.1915, Jerusalem, **PA/AA**, Türkei, 152. Dieckmann, "Die Hedschasbahn im Weltkriege…", p. 898. Dieckman, "Dr. h. c. Meissner Pascha…", p. 64.
66. Telegram from Wangenheim to Foreign Ministry, 10.11.1914, Tarabya, **PA/AA**, Türkei 152.
67. Pick, "Der deutsche Pionier …", p. 285.
68. Report by Construction and Management Adviser Allmarus, 19.10.1914, Berlin-Wilmersdorf, **PA/AA**, Türkei 152.
69. Dieckmann, "Die Hedschasbahn…", p. 898. See also Richard Hennig, "Die Mekkabahn und ihre Bedeutung für den Krieg", **Deutsche Strassen - und Kleinbahn - Zeitung**, Nr. 29 (15.7.1916), p. 348.
70. Dieckmann, "Suriye ve Lübnan Demiryolları", **Demiryollar Mecmuası**, no. 64 (1930), p. 194.
71. "Bahnbauten in der Türkei", **Kölnische Zeitung**, 5.2.1915. From Padel to Wangenheim, 1.2.1915, Damaskus, **PA/AA**, Türkei 152.
72. From Jaeckh to Foreign Affairs,13.5.1915, Berlin, **PA/AA**, Türkei 152.
73. From Chief of General Staff to Hollweg, 3.11.1915, Jaffa, **PA/AA**, Türkei 152.
74. Dieckmann, "Die Hedjasbahn und die syrischen Privatbahnen im Weltkiege und ihre gegenwaertige Lage", **Zwischen Kaukasus und Sınai**, vol. 2, (1922), p. 53. See also from Hardegg to Hollweg, 29.11.1915, Damascus, **PA/AA**, Türkei 152.
75. Cemal Pasha, p. 393; Dieckmann, "Die Hedschasbahn…", p. 898.

76. Dieckmann, "Die Hedjasbahn im Weltkriege...", p. 898. **Frankfurter Zeitung**, 26.5.1918. Dieckmann, "Kriegsbahnbauten im Nahen Osten, AfEW, 1942, p. 819.
77. Liman von Sanders, **Türkiye'de 5 Yıl**, İstanbul, 1968, pp. 45, 292, 293.
78. From Padel to Wangenheim, 1.2.1915, Damascus, **PA/AA**, Türkei 152.
79. Dieckmann, "Die Hedschasbahn...", pp. 895, 896.
80. Cemal Pasha, p. 394.
81. Ali Fuad Erden, **Birinci Dünya Harbinde Suriye Hatıraları**, vol. I, İstanbul, 1954, p. 252.
82. Sanders, p. 293.
83. Dieckmann, "Die Hedjasbahn im Weltkriege...", p. 896.
84. From Ranzi to Prime Minister Chudenitz, 14.4.1917, Damaskus, **HHStA**, Admin. Reg., F 19/33.
85. Report titled "Die ungarische Eisenbahnermission in Damaskus" from Ranzi to Pallavicini, 20.4.1917, Damascus, **HHStA**, Admin. Reg., F 19/33. Report titled "Rückreise von Angestellten der Hedjazbahn" from Pandini to Prime Ministry and Foreign Affairs, 12.4.1917, Aleppo, **HHStA**, Admin. Reg., F 19/33.
86. From Military Attaché Pomiankowski to Ministry of Defence, 16.4.1917, İstanbul, **HHStA**, Admin. Reg., F 19/33.
87. Report titled "Ungarische Angestellte der Hedschas-Bahn", from Pomiankowski to Ministry of Defence, 30.4.1917, İstanbul, **HHStA**, Admin. Reg., F 19/33.
88. For example report titled "Die ungarische Eisenbahnmission in Damaskus" from Trade Attaché of İstanbul to Prime Ministry and Ministry of Defence, 30.4.1917, İstanbul, **HHStA**, Admin. Reg., F 19/33.
89. Report titled "Waehrung der Interessen in den Dienst der türkischen Hedjazbahngetretenen ungarischen Staatsangehörigen" from Pandini to Prime Ministry and Foreign Affairs, 25.5.1917, İstanbul, **HHStA**, Admin. Reg., F 19/33.
90. Report titled "Die ungarische Eisenbahnermission in Damaskus" from Ranzi to Chudenitz, 1.5.1917, Damascus, **HHStA**, Admin. Reg., F 19/33.
91. Dieckmann, "Die Hedschasbahn...", p. 896.
92. Gülsoy, pp. 225, 226.
93. Interview with Abdullah by G. Antonius, **B.D.**, vol. 10/2, p. 832.
94. From Kitchener to W. Tyrrell, 26 April 1914, Kahire, **B.D.**, vol. 10/2, p. 831.
95. From Kitchener to Edward Grey, 21 March 1914, **B.D.**, vol. 10/2, p. 830.
96. From Kitchener to Edward Grey, 4 April 1914, **B.D.**, vol. 10/2, p. 830.
97. Dawn, pp. 11, 20, 21 et al.
98. Elie Kedouire, **England and the Middle East**, London, 1987, pp. 52, 53.
99. Isaiah Friedman, **British Pan-Arap Policy**, 1915–1922, New Brunswick and London, 2010, p. 25.
100. Fromkin, p. 174. Dawn, p. 28.
101. George Antonius, **The Arab Awakening, The Story of the Arab National Movement**, London, 1938, pp. 188, 190. Cemal Pasha had a book prepared

against the accusations that the judgements were unfair: **Osmanlı İmparatorluğu'nda Ayrılıkçı Arap Örgütleri, Âliye Divân-ı Harb-i Örfîsi,** İstanbul, 1993. In this book, he mentioned the argument that Arabian communities had specific goals, such as establishing the caliphate in Egypt under the rule of the British, getting Syria under the invasion of the French; see p. 11. For capital punishments and exiles, Kayalı, pp. 217–21 and Stoddard, pp. 124–9. The court was named after town of Aliyye, where the members gathered, Cemal Kutay, **Tarihte Türkler, Araplar, Hilâfet Meselesi,** İstanbul, 1970, p. 256.

102. William Cleveland, **Batı'ya Karşı İslam, Şekip Arslan'ın Mücadelesi,** İstanbul, 1991, p. 85.

103. Antonius, pp. 157, 158. Friedman, p. 47.

104. Fromkin, p. 185.

105. "...But, not beeing a perfect fool, I could see that if we won the war the promises to the Arabs were dead paper. Had I been an honourable aoviser I would have sent my men home, and not let them risk their lives for such stuff. Yet the Arab inspiration was our main tool in winning the Eastern war. So I assured them that England kept her word in letter and spirit. In this comfort they performed their fine things: but, of course, instead of being proud of what we did together, I was continually and bitterly ashamed", T. E. Lawrence, **Seven Pillars of Wisdom,** New York, 1926, pp. 275, 276.

106. Friedman, pp. 52–7.

107. Feridun Kandemir, **Medine Müdafaası,** İstanbul, 1991, pp. 455, 459, 460.

108. Stoddard, pp. 118, 121, 122.

109. Bayur, p. 240; Dawn, p. 34; Kandemir, p. 465.

110. Ali Fuad Erden, **Paris'ten Tîh Sahrasına,** Ankara, 1949, p. 59. Kayalı, pp. 211, 212.

111. Ömer Kürkçüoğlu, **Osmanlı Devleti'ne Karşı Arap Bağımsızlık Hareketi (1908–1918),** Ankara, 1982, pp. 115–20.

112. Briton Cooper Busch, **Britain, India, and the Arabs, 1914–1921,** Berkeley, Los Angeles, London, 1971, p. 171.

113. For the trouble of such division, **Erkân-ı Harb Binbaşısı Vecihi Bey'in Anıları, Filistin Ricatı,** İstanbul, 1993, pp. 19, 20.

114. Gülsoy, p. 230. Kandemir, p. 462.

115. Wallach also shared Lawrence's view, when saying: "Arab is a good fighter but a bad soldier", Jehuda L. Wallach, "Ein deutscher Gelehrter und Offizier aeussert sich im ersten Weltkrieg zur Frage des militaerischen Wertes der Araberstaemme", **Jahrbuch des Instituts für Deutsche Geschichte,** vol. 4 (1975), p. 474.

116. T.E. Lawrence, **Revolt in the Desert,** New York, 1927, pp. 66, 67.

117. Kandemir, pp. 56, 57.

118. İsmet Bozdağ, **Osmanlı'nın Son Kahramanları, Batı Trakya Türkleri ve Medine Müdafaası,** İstanbul, 1996, p. 76.

119. See examples from Lawrence' memories; Lawrence, **Revolt** ..., pp. 102, 103, 108, 129, 140, 141, 148, 149, 180, 232, 236, **Seven**..., pp. 244, 425, 431, 432, et al.

120. Fromkin, pp. 312. See also Friedman, pp. 64. 66. "Sharif Husayn's riot cost 11 million pounds for the British", Stoddard, p. 177.

121. Telegram from Waldburg to Foreign Ministry, 4.8.1917, Pera, **PA/AA**, Türkei 152.

122. For the reports of the experience of an engineer fixing the damaged trains, Atayman, pp. 73–99. Gülsoy, pp. 231–4, 241. Kuşçubaşı, p. 256:

123. **B.O.A., DH. EUM,** 4. Şb 1/46.

124. Nâci Kâşif Kıcıman, **Medine Müdafaası,** İstanbul, 1994, pp. 157, 158.

125. Ochsenwald, pp. 129, 130.

126. Joseph Pomiankowski, **Der Zusammenbruch des Ottomanischen Reiches, Erinnerungen an die Türkei aus der Zeit des Weltkrieges,** Zürich-Leipzig-Wien, 1928, p. 92. For detailed infromation, Dieckmann, "Die syrischen und palaestinischen Eisenbahnen im Kriege", **AfEW**, 1942.

127. Pönicke, p. 28.

128. Hejaz railway is included, the average for Ottoman Railways falls from 1,570 to 1,250, Hecker, pp. 774, 775.

129. For the resistance of caravan trade, Ortaylı's following analysis may give remarkable clues, İlber Ortaylı, "Devenin Taşıma Eğrisi Üzerine Bir Deneme", **Ankara Üniversitesi Siyasal Bilgiler Fakültesi Dergisi**, vol. 28, no. 1–2 (March-June 1973), especially pp. 186–8.

130. "Pilgrims come to the shop. Shop owners welcome them with respect. They serve them wonderful perfumes. Meccans make a living thanks to pilgrims. That is why the sharif asks advises us to trat pilgrims extremely well... " For further remarks, Evliya Çelebi, **Tam Metin Seyahatnâme**, vol. 9, İstanbul, 1996, p. 166 et al. For Wavell's interesting remarks on the same subject, Wavell, p. 64 et al. See also Faroqhi, **Hacılar**..., p. 200.

131. C. Snouck Hurgronje, **Mekka**, Haag, 1889, p. 38.

132. A. J. Wensinck, "Hacc", **İslâm Ansiklopedisi**, vol. 5/1, İstanbul, 1988, p. 13.

133. From Toncic to Aehrenthal, 30.3.1909, Jeddah, **HHStA**, Admin. Reg. F 19/18.

134. Ochsenwald, pp. 122, 123. For an exaggerated remark claiming that Bedouins supported the railway (for this very reason), Rudolf Reinhard, "Die Hedschasbahn", **Universum**, Nr. 5, vol. II, 1909, p. 926.

135. From Pallavicini to Aehrenthal, 6.3.1908, İstanbul, **HHStA**, Admin. Reg., F 19/18, pp. 30, 31.

136. Andre Raymond, **Arab Cities in the Ottoman Periode: Cairo, Syria, and the Maghreb**, Burlington, 2002, p. 27.

137. For the numbers indication how agricultural production grew bigger in 1894 thanks to Anatolian Railway, Zander, "Einwirkung...", pp. 24, 25.

138. For similar claims, from Hardegg to Hollweg 16.7.1914, Haifa, **PA/AA**, Türkei 152, p. 19; Auler p. 22. For agricultural wealth in ancient times, "Die Hedschaseisenbahn", **Der neue Orient**, Nr. 4 (12.1.1920).

139. Auler (1908), p. 59.

140. From Oppenheim to Bülow, 29.4.1903, Cairo, **PA/AA**, Türkei 152, pp. 15, 16.

141. Oppenheim's report, p. 15.

142. **B.O.A, Y. MTV**, 262/46.

143. Reply from Muslim Refugees Commission, 7 October 1907, **B.O.A., DH. MKT**, 1181/44.

144. **B.O.A, Y. A. Res**, 135/40. For a note emphasizing the settlement of immigrants along the railway, **B.O.A, Y. PRK. BŞK**, 78 /1.

145. Eugen Wirth, "Die Rolle tscherkessischer Wehrbauern bei der Wiederbe-siedlung von Steppe und Ödland im Osmanischen Reich," **Bustan**, Heft 1, (1963), p. 18. See also "Die Mekkabahn", **Archiv für Post und Telegraphie**, (1905), p. 782.

146. Ortaylı, "19. Yüzyıl...", p. 97.

147. Wirth, "Die Rolle tscherkessischer...", p. 18. See also F. Thiess, "Die Hedschasbahn", **Illustrierte Zeitung**, Nr. 3630 (23 January 1913), p. 204.

148. Schakib Arslan, "Um die Hedschaslinie", **Weser-Zeitung**, Nr. 343 (15.5.1918), p. 1.

149. **B.O.A, Y. PRK. SGE**, 10/61.

150. Kapp von Gülstein, 6.11.1905, **B.O.A., Y. PRK. TNF**, 8/35.

151. From Schroeder to Bülow, 26.6.1905, Beirut, **PA/AA**, Türkei 152, pp. 17, 18.

152. "Französische Eisenbahnprojekte in Syrien und Palaestina ", **DLZ**, 1.9.1912.

153. From Minister of Justice to Internal Affairs, 19 August 1910, **B.O.A., DH. İD**, 52/5. See also **B.O.A. MV**, 138/26.

154. From Pallavicini to Aehrenthal, 6.3.1908, İstanbul, **HHStA**, Admin. Reg., F 19/18, p. 23.

155. From Schroeder to Bülow, 26.6.1905, Beirut, **PA/AA**, Türkei 152, p. 19.

156. Joseph Maria Hunck, "Die Bahn der Pilger", **Handelsblatt**, Nr. 220, 14.11.1967, p. 22.

157. Thankmar von Münchhausen, "Eine Eisenbahn durch die Wüste Südjorda-niens", **Frankfurter Allgemeine Zeitung**, 28.5.1973.

158. From Schroeder to Bülow, 26.6.1905, Beirut, **PA/AA**, Türkei 152, p. 15. For the guest house, see also P. Dieckmann, "Nachricht für Reisende auf der Hedschasbahn", **Zeitschrift des Deutschen Palaestina-Vereins**, vol. 37 (1914).

159. **DLZ**, 15.6.1914, s. 54.

160. Office of Finance and Education of Council of State, **B.O.A., DH. İD**, 132/13.

161. Gülsoy, p. 248.

162. From Governor of Syria to Internal Affairs, 14 November 1911, **B.O.A., DH. İD**, 3/58.

163. Ochsenwald, pp. 135, 136.

164. Graves, p. 87.

165. Ochsenwald, p. 138.

166. "Die syrischen Eisenbahnen Damaskus-Hamah und die Hedschasbahn", **AfEW**, 1905.

167. From Governor of Beirut to Internal Affairs, 25 March 1909, **B.O.A., DH. İD**, 54/96.

BIBLIOGRAPHY

I-PRIMARY SOURCES

A. BAŞBAKANLIK OSMANLI ARŞİVİ

-YILDIZ TASNİFİ

Esas Ve Sadrazam Kâmil Paşa (Y. EE)
Hazine-i Hassa (Y. PRK. HH)
Kamil Paşa Evrakı (Y. EE. KP)
Komisyonlar Maruzatı (Y.PRK.KOM)
Maarif Nezareti Mevzuatı (Y. PRK. MF)
Mabeyn Erkanı ve Saray Görevlileri Maruzatı (Y. PRK. SGE)
Maliye Nezareti Maruzatı (Y.PRK.ML)
Maruzat Defterleri (Y. MRZ.d)
Mütenevvi Maruzat Evrakı (Y. MTV)
Perakende Evrakı Arzuhal Ve Jurnaller (Y. PRK. AZJ)
Perakende Evrakı Askerî Maruzat (Y.PRK.ASK)
Perakende Evrakı Mabeyn Başkitabeti (Y.PRK.BŞK)
Perakende Evrakı Elçilik, Şehbenderlik Ve Ataşemiliterlik (Y. PRK. EŞA)
Perakende Evrakı Tahrirat-ı Ecnebiye Ve Mabeyn Mütercimliği (Y. PRK. TKM)
Sıhhiye Nezareti Maruzâtı (Y. PRK. SH)
Sadaret Hususi Maruzat Evrakı (Y.A. Hus)
Sadaret Resmi Maruzat Evrakı (Y.A. Res)
Sadaret (Y. PRK. A)
Ticaret ve Nafıa Nezareti Maruzatı (Y. PRK. TNF)
Umumi (Y. PRK. UM)

-İRADELER TASNİFİ

Askeri (İ. AS)
Hususi (İ. Hus)

Mabeyn-i Hümayun (İ. MBH)
Maliye (İ. ML)
Meclis-i Mahsus (İ. MMS)
Meclis-i Umumi (İ: MLU)
Ticaret ve Nafıa (İ. TNF)

-DAHİLİYE NEZARETİ BELGELERİ

Emniyet-i Umumiye (DH.EUM)
İdarî Kısım Belgeleri (DH. İD)
Kalem-i Mahsus (DH. KMS)
Mektubi Kalemi (DH. MKT)
Muhaberât-ı Umumiye İdaresi Belgeleri (DH. MUİ)

-HARİCİYE NEZARETİ BELGELERİ

Siyasi Kısım (HR. SYS)
Tercüme Odası (HR. TO)

-DİĞER BELGELER

Meclis-i Vükela Mazbataları (MV)
Sadaret Mühimme Kalemi Evrakı (A. MKT. MHM)
Zaptiye Nezareti Belgeleri (Z. B)

B. POLITISCHES ARCHIV DES AUSWAERTIGEN AMTES
Türkei 152: Die Eisenbahnen in der asiatischen Türkei
Türkei 165: Arabien
Archiv der Kaiserlich Deutschen Botschaft in Konstantinopel

C. HAUS – HOF – UND STAATSARCHIV, WIEN
Administrative Registratur, F 19/18
Administrative Registratur, F 19/33
Botschaftsarchiv Konstantinopel

II- PRINTED PRIMARY SOURCES

Bagdad Railway Convention of March 5 1903, Presented to both Houses of Parliament by Command of His Majesty, May 1911, London (published by His Majesty's Stationery Office).
British Documents on Foreign Affairs, Reports and Papers from the Foreign Office Confidential Print, General eds K. Bourne and D. Cameron Watt, vol. 20, ed. D. Gillard, University Publications of America, 1985.
British Documents on the Origins of the War 1898–1914, eds G.P. Gooch and Harold Temperley, vol. III, London, 1928. vol. V, London 1928. vol. X/I, London, 1936. vol. X/II, London, 1938.
German Diplomatic Documents, 1871–1914, Selected and translated by E.T.S. Dugdale, vol. II, London, 1929, vol. III, London, 1930.

Die Grosse Politik der Europaeischen Kabinette 1871–1914, Sammlung der Diplomatischen Akten des Auswaertigen Amtes, Im Auftrage Auswaertigen Amtes, eds Johannes Lepsius, Albrecht Mendelssohn Bartholdy, Friedrich Timme. Band 14/2, Weltpolitische Rivalitaeten, Berlin 1924.

Hicaz Demiryolu, Hicaz Demiryolu'nun Vâridât ve Mesârifi ve Terâkki-i İnşaatı ile Hattın Ahvâl-i Umumiyyesi Hakkında Malumât-ı İhsâiyye ve İzahât-ı Lazımeyi Muhtevidir, Year 5, 1330, İstanbul 1334.

Meclisi Mebusan Zabıt Ceridesi, TBMM Basımevi, Session 35, vol. 2, Ankara 1982.

The Parliamentary Debates, House of Commons, fifth series, vol. 59, London, 1914.

III- SECONDARY SOURCES

ABDULLAH: Memoirs of King Abdullah of Transjordan, ed. Philip P. Graves, London, 1950.

ADHAM M. Fahm: "Between Mystical and Military: The Architecture of the Hejaz railway (1900–1908)", Essays in Honour of Aptullah Kuran, ed. Ç. Kafescioğlu-L. Thys-Şenocak, İstanbul, 1999.

AHMAD, Feroz: İttihat ve Terakki, (1908–1914), İstanbul, 1984.

AHMAD, Feroz: İttihatçılıktan Kemalizme, İstanbul, 1985.

AITCHISON G.U., B.G.S.: A Collection of Treaties, Engagaments and Sanads, Relating to India and Neighbouring Countries, Delhi, 1933.

AKARLI, Engin Deniz: Belgelerle Tanzimat, Osmanlı Sadrazamlarından Ali ve Fuat Paşaların Siyasi Vasiyetnameleri, İstanbul, 1978.

AKARLI, Engin Deniz: "II. Abdülhamid (1876–1909)", Tanzimattan Cumhuriyete Türkiye Ansiklopedisi, vol. 5, İstanbul, 1985.

AKPINAR, Alişan: Osmanlı Devletinde Aşiret Mektebi, İstanbul, 1997.

AKŞİN, Sina: Jön Türkler ve İttihat ve Terakki, İstanbul, 1980.

AKYILDIZ, Ali: "Osmanlı Anadolusunda İlk Demiryolu: İzmir – Aydın Hattı (1856–1866)", ÇYO, ed. E. İhsanoğlu, M. Kaçar, İstanbul, 1995.

«Die Anatolischen Eisenbahnen in den Jahren 1905 und 1906», AfEW, 1908.

ANDERSON, M. S.: The Eastern Question, 1774–1923, A Study in International Relations, New York, 1966.

ANTONIUS, George: The Arab Awakening, The Story of the Arab National Movement, London, 1938.

ARAR, İsmail: «Yakın Tarihimizden Portreler, Macera Dolu Bir Hayat: Vehip Paşa», Tarih ve Toplum, no. 48, 1987.

ARHANGELOS, Gavriel: Anadolu Osmanlı Demiryolu ve Bağdat Demiryolu Şirket-i Osmaniye İdaresi'nin İçyüzü, İstanbul, 1327.

ARTUK, Cevriye: "Hicaz Demiryolu Madalyaları", VII. Türk Tarih Kongresi, Kongreye Sunulan Bildiriler, (25–29 Eylül 1970), vol. II, Ankara, 1963.

ARTUK, İbrahim: «Hicaz Demiryolu'nun Yapılması ve Bu Münasebetle Basılan Madalyalar», İstanbul Arkeoloji Müzeleri Yıllığı, no. 11–12, (1964).

ATALAR, Münir: Osmanlı Devletinde Surre-i Hümâyun ve Surre Alayları, Ankara, 1991.

ATAYMAN, Mustafa Şevki: Bir İnşaat Mühendisinin Anıları 1897–1918, İstanbul, 1984.

AULER Pascha: «Besprechung», AfEW, 1907.

AULER Pascha: «Besprechung», AfEW, 1908.

AULER Pascha: Die Hedschasbahn, Auf Grund einer Besichtigungsreise und nach amtlichen Quellen, Gotha, 1906.

AULER Pascha: Die Hedschasbahn, II. Teil: Ma'ân bis El Ula, Auf Grund einer zweiten Besichtigungsreise und nach amtlichen Quellen, Gotha, 1908.

AYDIN Ayşe H. (ed.): Osmanlı İmparatorluğu'nda Ayrılıkçı Arap Örgütleri, Âliye Divân-ı Harb-i Örfîsi, İstanbul, 1993.

"Bahnbauten in der Türkei", Kölnische Zeitung, 5.2.1915.

BARBIR Karl K.: "Bellek, Miras ve Tarih: Arap Dünyasında Osmanlı Mirası", İmparatorluk Mirası, Balkanlar'da ve Ortadoğu'da Osmanlı Damgası, ed. L. Carl Brown, İstanbul, 2000.

BAYUR, Yusuf Hikmet: Türk İnkılâbı Tarihi, vol. III, Part III, Ankara, 1991.

BECKER, Kurt: "Die Hedschasbahn", Orient, no. 5 (April 1963).

«Die Beduinen im Bereiche der Hedschasbahn», Hannoverscher Courier, 21.12.1910.

BERDROW, Wilhelm: "Die Hedschasbahn", ZdVDEV, Nr. 97, 15.12.1906.

BERDROW, Wilhelm: "Die Hedschasbahn", ZdVDEV, Nr. 83, 15.12.1906.

"Bir Demiryolu Grevi", Tarih ve Toplum, no. 5 (May 1984).

BLAISDELL, Donald C.: European Financial Control in the Ottoman Empire, A Study of the Establishment, Activities, and Significance of the Administration of the Ottoman Public Debt, New York, 1929.

BLANCKENHORN, Max: «Die Hedschasbahn», Geographische Zeitschrift, Nr. XVIII, 1912.

BLANCKENHORN, Max: «Die Hedschaz-Bahn, Auf Grund einer Reisestudien», Zeitschrift der Gesellschaft für Erdkunde zu Berlin, no. 4,1907.

BONIN, Charles Eude: «Le chemin de fer du Hedjaz», Annales de Geographie, 1909.

BOZARSLAN, Mehmet Emin: «Hicaz Hamidiye Demiryolu İnşaatı», 20. Yüzyıl Tarihi, vol. I, İstanbul, 1970.

BOZDAĞ, İsmet: Osmanlı'nın Son Kahramanları, Batı Trakya Türkleri ve Medine Müdafaası, İstanbul, 1996.

BOZKURT, Gülnihal: Gayrimüslim Osmanlı Vatandaşlarının Hukuki Durumu (1839–1914), Ankara, 1989.

BUSCH, Briton Cooper: Britain India and the Arabs, 1914–1921, Berkeley, Los Angeles, London, 1971.

CARMEL, Alex: "Die deutsche Palaestinapolitik 1871–1914", Jahrbuch des Instituts für Deutsche Geschichte, 4. Band (1975).

CEMAL Paşa: Hatıralar, İstanbul, 1977.

CEMAL Paşa: Osmanlı İmparatorluğu'nda Ayrılıkçı Arap Örgütleri, Âliye Divân-ı Harb-i Örfîsi, İstanbul, 1993.

ÇETİNYALÇIN, İsmet: «Liyakat Madalyası», VIII. Türk Tarih Kongresi, Kongreye Sunulan Bildiriler (11–15 Ekim 1976), vol. 3, Ankara, 1983.

CHESNEY, Francis Rawdon: Narrative of the Euphrates Expedition, (Carried on by Order of the British Government during the Years 1835, 1836, and 1837), London, 1868.

CLAYTON, G. D.: Britain and the Eastern Question: Missolonghi to Gallipoli, London, 1971.

CLEVELAND, William: Batı'ya Karşı İslam, Şekip Arslan'ın Mücadelesi, İstanbul, 1991.

ÇULCU Murat (ed.): Erkân-ı Harb Binbaşısı Vecihi Bey'in Anıları, Filistin Ricatı, İstanbul, 1993.

DAWN, C. Ernest: From Ottomanism to Arabism, Urbana, Chicago London, 1973.

DEHN, Paul: Deutschland und die Orientbahnen, München, 1883.

DENICKE: «Die Hedschasbahn», Glasers Annalen für Geverbe und Bauwessen, Nr. 760, 15.2.1909.

DERİNGİL, Selim: "Osmanlı İmparatorluğu'nda 'Geleneğin İcadı', 'Muhayyel Cemaat' ('Tasarımlanmış Topluluk') ve Panislamizm", Toplum ve Bilim, no. 54/55.(Yaz/Güz 1991).

DIECKMANN, P.: «Die Bautaetigkeit im Gebiet der Hedschasbahn», ZdVDEV, Nr. 74, 25.9.1912.

DIECKMANN, P.: «Die Pilgerbeförderung auf der Hedjazbahn», ZdVDEV, Nr. 33 (30.4.1913).

DIECKMANN, P.: «Nachricht für Reisende auf der Hedschasbahn», Zeitschrif des Deutschen Palaestina Vereins, Band 37 (1914).

DIECKMANN, P.: «Ergebnisse der Regenmessung im Hedschazbahngebiet, Winter 1913/14», Zeitschrift des Deutschen Palaestina Vereins, vol. 37, 1914.

DIECKMANN, P.: «Die Zweiglinie Affule-Jerusalem der Hedschazbahn», Zeitschrift des Deutschen Palaestina Vereins, Band 37 (1914).

DIECKMANN, P.: «Die muslimische Pilgerbewegung nach Mekka und Medina», DLZ, Nr. 21/22, 1.11.1915.

DIECKMANN, P.: «Die Hedschasbahn im Weltkriege und ihre gegenwaertige Lage», ZdVDEV, Nr. 48 (1.12.1921).

DIECKMANN, P.: «Die Hedjasbahn und die syrischen Privatbahnen im Weltkriege und ihre gegenwaertige Lage», Zwischen Kaukasus und Sinai, vol. 2 (1922).

DIECKMANN, P.: "Suriye ve Lübnan Demiryolları", Demiryollar Mecmuası, no. 64, (1930).

DIECKMANN, P.: "Wierderherstellung der Hedschazbahn bis Medina und Weiterbau bis Mekka und Dschidda", Petermanns Mitteilungen, Nr. 82, Gotha, 1936.

DIECKMANN, P.: «Dr h.c. Meissner Pascha, der Erbauer der Hedjasbahn», Zeitung des Vereins Mittereuropaeischer Eisenbahnverwaltungen, (1940).

DIECKMANN, P.: «Kriegsbahnbauten im Nahen Osten», AfEW, 1942.

DIECKMANN, P.: «Die syrischen und palaestinischen Eisenbahnen im Kriege», AfEW, 1942.

DIECKMANN, P.: «Die verkehrsgeographische Bedeutung der Vierlaenderecke bei Akaba», AfEW, 1943.

DİNÇER, Celal: «Osmanlı Vezirlerinden Hasan Fehmi Paşa'nın, Anadolu'nun Bayındırlık İşlerine Dair Hazırladığı Layiha», T.T.K. Belgeler Dergisi, vol. V-VIII, (offprint from no. 9–12), Ankara, 1972.

EARLE Edward Mead: Turkey, The Great Powers and the Bagdad Railway, A Study in Imperialism, New York, 1923.

«Das Eigentumsrecht an der Hedjazbahn», ZdVDEV, Nr. 41, 8.10.1931.

"Eisenbahn Haifa Damascus", **Wiener Politische Correspondenz**,14.3.1902.
«Die Eisenbahnen der Türkei im Jahre 1904», **AfEW**, 1907.
«Die Eisenbahnen der Türkei im Jahre 1905», **AfEW**, 1908.
«Die Eisenbahnen der Türkei im Jahre 1906», **AfEW**, 1909.
«Die Eisenbahnen der Türkei im Jahre 1907», **AfEW**, 1910.
«Die Eisenbahnen der Türkei im Jahre 1909», **AfEW**, 1912.
«Die Eisenbahnen der Türkei im Jahre 1910», **AfEW**, 1912.
«Die Eisenbahnen der Türkei im Jahre 1911», **AfEW**, 1913.
"Die eisenbahnlose Zone der Englaender", **Neue Preusische Zeitung**, 8.4.1908.
ELDEM, Vedat: **Osmanlı İmparatorluğu'nun İktisadi Şartları Hakkında Bir Tetkik**, Ankara, 1970.
ELEFTERÍADES, Eleuthere: **Les chemins de fer en Syrie et au Liban**, Beirut, 1944.
ENGELBRECHTEN, C. A.: **Kaiser Wilhelms Orientreise und deren Bedeutung für den deutschen Handel**, Berlin, 1890.
ENGİN, Vahdettin: **Rumeli Demiryolları**, İstanbul, 1993.
«Die Entwicklung der Eisenbahnen in der europaeischen und asiatischen Türkei, insbesondere in Syrien», **ZdVDEV**, Nr. 57, 22.7.1908.
ERASLAN, Cezmi: **II. Abdülhamid ve İslam Birliği**, İstanbul, 1992.
ERDEN, Ali Fuad: **Birinci Dünya Harbinde Suriye Hatıraları**, İstanbul, 1954.
ERDEN, Ali Fuad: **Paris'ten Tîh Sahrasına**, Ankara, 1949.
ERGÜL, Cevdet: **II. Abdülhamid'in Doğu Politikası ve Hamidiye Alayları**, İzmir, 1997.
ERKİN, Osman: «Demiryolu Tarihçesinden: Hicaz Demiryolu», **Demiryollar Dergisi**, no. 269, (1948).
«Eröffnung einer Zweiglinie der Hedschasbahn», **AfEW**, 1913.
«Die Eröffnung der Bahnstrecke Haiffa – Bell – ede – Scheikh – Akko.» **DLZ**, Nr. 22, 15.11.1913.
EVLİYA, Çelebi: **Tam Metin Seyahatnâme**, vol. 9, İstanbul, 1996.
FAROQHİ, Suraiya: **Herrscher über Mekka, Die Geschichte der Pilgarfahrt**, München und Zürich, 1990.
FAROQHİ, Suraiya: **Hacılar ve Sultanlar (1517–1683)**, İstanbul, 1995.
FELDMANN, Wilhelm: «Bei Meissner Pascha.», **Berliner Tageblatt**, Nr. 31, 18.1.1917.
FIEDLER, Ulrich: **Der Bedeutungswandel der Hedschasbahn; Eine historisch geographische Untersuchung**, Berlin, 1984.
FISCHER, Fritz: **Griff nach der Weltmacht**, Düsseldorf, 1964.
«Französische Eisenbahnprojekte in Syrien und Palaestina», **DLZ**, 1.9.1912.
"Französische Förderungen an die Türkei", **Frankfurter Zeitung**, 26.3.1905.
"Französische Plaene in Syrien und Palaestina I", **Schlesische Zeitung**, 29.10.1913.
"Französische Plaene in Syrien und Palaestina II", **Schlesische Zeitung**, 30.10.1913.
FRIEDMAN, Isaiah: **British Pan-Arap Policy, 1915–1922**, New Brunswick and London, 2010.
FROMKİN, David: **A Peace to End All Peace: Creating the Modern Middle East, 1914–1922**, New York, 1989.
FULTON, L. Bruce: "Fransa ve Osmanlı İmparatorluğu'nun Sonu", **Osmanlı İmparatorluğu'nun Sonu ve Büyük Güçler**, ed. Marian Kent, İstanbul, 1999.

280   THE HEJAZ RAILWAY AND THE OTTOMAN EMPIRE

GEDİKL, İ Fethi: "Midhat Paşa'nın Suriye Layihası", Dîvân İlmî Araştırmalar, Year 4, no. 7 (1999/2).

GEORGEON, François: «II. Abdülhamid ve İslam», Tarih ve Toplum, no. 112, April, 1993.

GEORGEON, François: «Son Canlanış (1878–1908)», Osmanlı İmparatorluğu Tarihi II, ed. Robert Mantran, İstanbul, 1995.

GEORGEON, François: Sultan Abdülhamid, İstanbul, 2006.

"Die geplante Mekka – Eisenbahn", Frankfurter Zeitung, 10.12.1900.

Von der GOLTZ, Freiherr Colmar: Denkwürdigkeiten, bearbeitet und herausgeg. von der Goltz Friedrich Freiherr und Wolfgang Förster, Berlin, 1932.

GÖÇEK, Fatma Müge and ÖZYÜKSEL, Murat: "The Ottoman Empire's Negotiation of Western Liberal Imperialism", Liberal Imperialism in Europe, ed. M.P. Fitzpatrick, New York, 2012.

GRAVES, Philip P.: Briton and Türk, London and Melbourne, 1941.

"Great Turkish Railway", The Globe, 22.7.1908.

GRECE, Michel de: II. Abdülhamid, Yıldız Sarayı'nda 30 Yıl, İstanbul, 1992.

GROTHE, Hugo: Meine Studienreise durch Vorderasien (Kleinasien, Mesopotamien u. Persien), 1906 and 1907, Halle, 1908.

GROTHE, Hugo: Die Bagdadbahn und das schwaebische Bauernelement in Transkaukasien und Palaestine, Gedanken zur Kolonisation Mesopotamiens, München, 1902.

GROTHE, Hugo: Die asiatische Türkei und die deutschen Interessen, Halle, 1913.

GUBOĞLU, Mihail P.: "Osmanlı İmparatorluğu'nda Karadeniz Tuna Kanalı Projeleri (1836–1876) ve Boğazköy – Köstence Arasında İlk Demiryolu İnşası (1855–1860), ÇYO, ed. E. İhsanoğlu, M. Kaçar, İstanbul, 1995.

GUTHE, Hermann: Die Hedschasbahn von Damaskus nach Medina, Ihr Bau und ihre Bedeutung, Leipzig, 1917.

GÜLSOY, Ufuk: Hicaz Demiryolu, İstanbul, 1994.

GÜLSOY, Ufuk – OCHSENWALD William: "Hicaz Demiryolu", Diyanet İslâm Ansiklopedisi, vol. 17, İstanbul, 1998.

GÜNERGÜN, Feza: "Osmanlı Devleti'nde Buharlı Tramvay İşletme Teşebbüsleri", ÇYO, ed. E. İhsanoğlu, M. Kaçar, İstanbul, 1995.

GÜRSEL, Seyfettin: "1838 Ticaret Antlaşması Üzerine", Yapıt, no. 10 (April–May, 1985).

GÜZEL, Şehmus: "Anadolu – Bağdat Demiryolu Grevi", Tanzimat'tan Cumhuriyet'e Türkiye Ansiklopedisi, vol. III (1985).

«Der Hafen der Mekkabahn», Deutsche Rundschau, Year 31 (1909).

HALLGARTEN, George W. F.: Imperialismus vor 1914, Die soziologischen Grundlagen der Aussenpolitik europaeischer Grossmaechte vor dem ersten Weltkrieg, München, 1963.

«Die Hamidie-Eisenbahn im Hedschas», Illustrierte Zeitung, Nr: 3168, (17.3.1904).

HARTIG, Paul: "Die Bagdadbahn-Idee und Verwirklichung", Zeitschrift für Geopolitik, Heidelberg, Berlin, 1940.

HARTMANN, Martin: «Das Bahnnetz Mittelsyriens», Zeitschrift des Deutschen Palaestina Vereins, Band 17, (1894).

HARTMANN, Martin: «Die Mekkabahn», Orientalistische Litteratur-Zeitung, Nr. 1, 15.1.1908.
HARTMANN, Martin: «Die Mekkabahn, ihre Aussichten und ihre Bedeutung für den Islam», Asien, Heft 10, (1912).
HARTMANN, Martin: Reisebriefe aus Syrien, Berlin, 1913.
HASLIP, Joan: The Sultan, The Life of Abdul Hamid II, London, 1958.
HASLIP, Joan: II. Abdülhamid, İstanbul, 1998.
HECKER, M.: "Die Eisenbahnen in der asiatischen Türkei", AfEW (pp. 744–800, 1057–1087, 1283–1321, 1539–1584), 1914.
«Die Hedschasbahn», AfEW, 1916.
«Die Hedschasbahn», ZdVDEV, Nr. 16, 17.4.1924.
"Die Hedjaz Eisenbahn", Mitteilungen der k.k. geographischen Gescllschaft in Wien, XLVIII (1904).
"The Hedjaz Railway, A Record of Jobbery", Times of India, 11.3.1907.
"Hedschas – und Bagdadbahn", Alldeutsche Blaetter, Nr. 49, 8.12.1906.
"Die Hedschasbahn (Chemin de fer Hamidie'du Hedjaz)", ZdVDEV, no. 89, (12.11.1904).
"Die Hedschasbahn", Die Umschau, Nr. 28 (9.7.1910).
"Die Hedschasbahn", DLZ, Nr. 6, 1.5.1912.
"Von der Hedschasbahn", DLZ, Nr. 12, 15.6.1914.
"Die Hedschasbahn", AfEW, 1916.
«Die Hedschasbahn, der Islam und Englands Stellung dazu», Die Grenzboten, Nr. 38, 17. 9. 1908.
"Die Hedschaseisenbahn", Der neue Orient, Nr. 4 (12.1.1920).
"Die Hedschasbahn" Gaea, Natur und Leben, Nr. 10 (1907).
«Die Hedschasbahn», Geographische Zeitschrift, Year 13, Nr. 3 (1907).
HEIGL, Peter: "Deutsche Bahnbauingenieure bei den Bauarbeiten der Hedjaz – und Bagdadbahn", Bagdad – und Hedjazbahn, Deutsche Eisenbahngeschichte im Vorderen Orient, ed. Jürgen Franzke, Nürnberg, 2003.
HEIGL, Peter: Schotter für die Wüste, Die Bagdadbahn und ihre deutschen Bauingenieure, Nürnberg, 2004.
HEINTZE, W.: "Eisenbahnen in der Türkei", Mitteilungen der geographischen Gesellchaft in Hamburg, vol. XI, 1896.
HELL, J.: "Stambul und Mekka", Erlanger Aufsaetze aus ernster Zeit (1917).
HELLER, Joseph: British Policy towards the Ottoman Empire 1908–1914, London, 1983.
HELLFERICH, Karl: Georg von Siemens, Ein Lebensbild aus Deutschlands grosser Zeit, Band III, Berlin, 1923.
HELLFERICH, Karl: Die deutsche Türkenpolitik, Berlin, 1921.
HENNIG, Richard Fridenau: «Das Projekt der transarabischen Bahn», DLZ, Nr. 17, 15.10.1912.
HENNIG, Richard Fridenau: Die Hauptwege des Weltverkehrs, Jena, 1913.
HENNIG, Richard Fridenau: «Die Mekkabahn und ihre Bedeutung für den Krieg», Deutsche Strassen – und Kleinbahn – Zeitung, Nr. 29 (15.7.1916).
HENNIG, Richard Fridenau: Die deutschen Bahnbauten in der Türkei, Leipzig, 1915.
HEUER, O.: «Die Hedschasbahn», DLZ, Nr. 5, 1.3.1913.

HIRSCH, Ernst E.: Anılarım; Kayzer Dönemi, Weimar Cumhuriyeti, Atatürk Ülkesi, Ankara, 1997.

Historical Section of the Foreign Office: Persian Gulf, vol. XIII, London, 1920.

HOPKIRK, Peter: On secret Service East of Constantinople: The plot to bring down the British Empire, London, 1994.

HUNCK, Joseph Maria: "Die Bahn der Pilger", Handelsblatt, Nr. 220, 14.11.1967.

HURGRONJE, C. Snouck: Mekka, Haag, 1889.

HÜLAGÜ, Metin: Pan-İslamist Faaliyetler, İstanbul, 1994.

HÜLAGÜ, Metin: Bir Umudun İnşası, Hicaz Demiryolu, İstanbul, 2008.

IMBERT, Paul: Osmanlı İmparatorluğu'nda Yenileşme Hareketleri, Türkiye'-nin Meseleleri, İstanbul, 1981.

INSHAULLAH, Muhammad: The History of the Hamidia Hedjaz Railway Project, printed at the Central Printing Works, Lahore, 1908.

İRTEM, Süleyman Kâni: Osmanlı Devleti'nin Mısır, Yemen, Hicaz Meselesi, İstanbul, 1999.

İRTEM, Süleyman Kâni: Şark Meselesi, Osmanlı'nın Sömürgeleşme Tarihi, İstanbul, 1999.

"Izzat Pasha and The Hedjaz Railway", The Times, 3.9.1908.

JERUSSALIMSKI, A.S.: Die Aussenpolitik und Diplomatie des deutschen Imperialismus Ende des 19. Jahrhunderts, Berlin, 1954.

KADRİ, Hüseyin Kazım: Balkanlardan Hicaza İmparatorluğun Tasfiyesi, İstanbul, 1992.

KANDEMİR, Feridun: Medine Müdafaası, İstanbul, 1991.

KARAL, Enver Ziya: Osmanlı Tarihi, vol. 8, Ankara, 1988.

KARKAR, Yaqub N.: Railway Development in the Ottoman Empire 1856–1914, New York, 1972.

KARPAT, Kemâl H.: "Pan-İslamizm ve İkinci Abdülhamid: Yanlış Bir Görüşün Düzeltilmesi", Türk Dünyasını Araştırmaları Dergisi, no. 47 (Haziran 1987).

KAYALI, Hasan: Jön Türkler ve Araplar, Osmanlıcılık, Erken Arap Milliyetçiliği ve İslamcılık (1908–1918), İstanbul, 1998.

KAYNAK, Muhteşem: «Osmanlı Ekonomisinin Dünya Ekonomisine Eklemlenme Sürecinde Osmanlı Demiryollarına Bir Bakış», Yapıt, no. 5 (1984).

KEDOURIE, Elie: England and the Middle East, The Destruction of the Ottoman Empire 1914–1921, London, 1987.

KENT, Marian: "Great Britain and the End of the Ottoman Empire, 1900–23", The Great Powers and the End of the Ottoman Empire, ed. M. Kent, London, 1984.

KICIMAN, Nâci Kâşif: Medine Müdafaası, Hicaz Bizden Nasıl Ayrıldı?, İstanbul, 1994.

KING, R.: "The Pilgrimage to Mecca: Some geographical and historical Aspects", Erdkunde, Nr. 26, (1972).

KLEIST, H.von: "Die Hedjasbahn", Asien, Heft 6 (Mart 1906).

KLEIST, H.von: "Die Hedjasbahn", Globus-Illustrierte Zeitschrift für Laender und Völkerkunde, vol. 89, Nr. 11, 22.4.1906.

KLEIST, H.von: «Die Hedschasbahn», Geographische Zeitschrift, vol. 3, 1907.

KOBER, Leopold: «Im nördlichen Hedschas», Hochland (February, 1912).

KOCHWASSER, Friedrich: «Der Bau der Bagdadbahn und die deutsche Orientpolitik», Deutsch-Türkische Gesellschaft E.V., Bonn Mitteilungen (June, 1975).

KODAMAN, Bayram: Sultan II. Abdülhamid Devri Doğu Anadolu Politikası, Ankara, 1987.

KODAMAN, Bayram: «II. Abdülhamid ve Aşiret Mektebi», Türk Kültürü Araştırmaları, XV/1–2, (1976).

KOLOĞLU, Orhan: Abdülhamit Gerçeği, İstanbul, 1987.

KOLOĞLU, Orhan: Bedevi, Lawrens, Arap, Türk, İstanbul, 1993.

KOLOĞLU, Orhan: "Dünya Siyaseti ve İslâm Birliği", Tarih ve Toplum, no. 83, (November 1990).

KOLOĞLU, Orhan: "Hicaz Demiryolu (1900–1908) Amacı, Finansmanı, Sonucu", ÇYO, ed. E. İhsanoğlu, M. Kaçar, İstanbul, 1995.

KOLOĞLU, Orhan: "Muktesid Musa'ya Göre Hicaz Demiryolu'nun Amacı", Toplum ve Ekonomi, no. 10 (July 1997).

KOLOĞLU, Orhan: "Medeniyet Götürmek İçin Yaptırılan Hicaz Demiryolu", Tarih ve Toplum, no. 184 (April 1999).

KURMUŞ, Orhan: Emperyalizmin Türkiye'ye Girişi, İstanbul, 1977.

KURŞUN, Zekeriya, "Şekib Arslan'ın Bazı Mektupları ve İttihadçılar ile İlişkileri", İstanbul Üniversitesi Edebiyat Fakültesi Tarih Enstitüsü Dergisi, no. 15, 1995–1997.

KUŞÇUBAŞI, Eşref: Hayber'de Türk Cengi, İstanbul, 1997.

KUTAY, Cemal: Tarihte Türkler, Araplar, Hilâfet Meselesi, İstanbul, 1970.

KÜTÜKOĞLU, Mübahat S.: Osmanlı-İngiliz İktisadi Münasebetleri II (1838–1850), İstanbul, 1976.

KÜRKÇÜOĞLU, Ömer: Osmanlı Devleti'ne Karşı Arap Bağımsızlık Hareketi (1908–1918), Ankara, 1982.

LANDAU, Jacob M.: The Hejaz Railway and the Muslim Pilgrimage, A Case of Ottoman Political Propaganda, Detroit, 1971.

LAWRENCE, T.E.: Seven Pillars of Wisdom, New York, 1926.

LAWRENCE, T. E.: Revolt in the Desert, New York, 1927.

LEVY, P.: «Die Betriebsmittel der Hedjazbahn», Organ für die Fortschritte des Eisenbahnwesens in technischer Beziehung, Nr. 6 (15.3.1911).

LINDOW, Erich: Freiherr Marschall von Bieberstein als Botschafter in Konstantinopel, 1897–1912, Danzig, 1934.

LORCH, Fritz.: «Die Eisenbahn Jaffa – Port Said», DLZ, Nr. 11, 15.7.1912.

LOWE, John: The Great Powers,Imperialism and the German Problem, 1865–1925, London and New York, 1981.

LUXEMBURG, Rosa: «Emperyalizmin Mısır ve Osmanlı İmparatorluğu'na Girişi», in Rathmann, Berlin-Bağdat Alman Emperyalizminin Türkiye'ye Girişi, ed. Ragıp Zarakolu, İstanbul, 1982.

LUXEMBURG, Rosa, Liebknecht Karl, Mehring Franz: The Crisis in the German Social Democracy, New York, 1818.

MAALOUF, Amin: Semerkand, İstanbul, 1998.

MAHMUT, Şevket Paşa: Mahmut Şevket Paşa'nın Günlüğü, İstanbul, 1988.

MANSFIELD, Peter: The Ottoman Empire and its Successors, London and Basingstoke, 1973.

MANZENREITER, Johann: **Die Bagdadbahn, Als Beispiel für die Entstehung des Finanzimperialismus in Europa** (1872–1903), Bochum, 1982.

MARDİN, Şerif: «İslamcılık», **Türkiye'de Din ve Siyaset**, ed. M. Türköne – T. Önder, İstanbul, 1991.

MEHRMANN, Karl Coblenz: **Der diplomatische Krieg in Vorderasien; Unter besonderer Berücksichtigung der Geschichte der Bagdadbahn**, Dresden, 1916.

MEJCHER, Helmut: «Die Bagdadbahn, Als Instrument deutschen wirtschaftlichen Einflusses im Osmanischen Reich», **Geschichte und Gesellschaft**, Nr 1–4, (1975).

"Mekkabahn", **Wiener Politische Correspondenz**, 9.1.1902.

«Die Mekkabahn», **Die Reform, Illustrieste Zeitschrift des Weltverkehrs**, Wien, 1903.

«Die Mekkabahn», **Die Reform, Internationales Verkehrsorgan**, 4. Jahrgang, 23. Heft (Erstes Augustheft 1903).

«Die Mekkabahn», **Archiv für Post und Telegraphie** (December, 1905).

MIJATOVICH, Chedo: «Abd ul Hamid», **Die Zukunft**, Nr. 47, 22.8.1908.

MOHR, Anton: **Der Kampf um Türkisch Asien. Die Bagdadbahn**, Meissen, n.a.

MOMMSEN, W.J.: **Europaeische Finanzimperialismus vor 1914, Ein Beitrag zu einer pluralistischen Theorie des Imperialismus**, Göttingen, 1979.

MORAWITZ, Charles: **Türkiye Maliyesi**, Maliye Bakanlığı Tetkik Kurulu Yayını, no. 1978–188, Ankara, 1978.

MORITZ, B.: "Der Hedjaz und die Strasse von Mekka nach Medina", **Zeitschrift der Gesellschaft für Erdkunde zu Berlin**, vol. 25, (1890).

MUSIL, Alois: **Im nördlichen Hegaz, Vorbericht über die Forschungsreise** 1910, Wien 1911.

MUSIL, Alois: **The Northern Hegaz**, New York, 1926.

MÜHLMANN, C.: «Die deutschen Bahnunterehmungen in der asiatischen Türkei 1888–1914», **Weltwirtschaftliches Archiv**, vol. 24, Jena, 1926.

MÜNCHHAUSEN, Thankmar von: "Eine Eisenbahn durch die Wüste Südjordaniens", **Frankfurter Allgemeine Zeitung**, 28.5.1973.

MYGIND, Eduard: **Vom Bosporus zum Sinai, Erinnerungen an die Einweihung der Hamidie Pilgerbahn des Hedjas (Teilstrecke Damaskus – Maan)**, Leipzig und Konstantinopol, 1905.

MYGIND, Eduard: **Syrien und die türkische Mekkapilgerbahn**, Halle, 1906.

NADİR, Tevfik: **Tehlikenin Büyüğü, Uyanalım**, İstanbul, 1324.

"Die neue Mekka-Bahn", **Die Katholischen Missionen**, Nr. 10 (1906/07).

«Die neuen Eisenbahn in Syrien», **Globus**, 68, 1895.

«Die neuen Eisenbahnen in der Türkei», **DLZ**, Nr. 23/24, 1.12.1915.

NOVIÇEV, A.D.: **Osmanlı İmparatorluğu'nun Yarı Sömürgeleşmesi**, Ankara, 1979.

NUMAN, İbrahim: «Hamidiye Hicaz Demiryolu», **Lâle**, no. 7, December 1990.

OCHSENWALD, William: "The Financing of the Hijaz Railroad", **Die Welt des Islams**, no. 14, Leiden, 1973.

OCHSENWALD, William: **The Hijaz Railroad**, University Press of Virginia, 1980.

OCHSENWALD, William: "Modern Ortadoğu'da İslâm ve Osmanlı Mirası", **İmparatorluk Mirası, Balkanlar'da ve Ortadoğu'da Osmanlı Damgası**, ed. L. Carl Brown, İstanbul, 2000.

OCHSENWALD, William: Religion, Society and The State in Arabia, Ohio, 1984.

ONUR, Ahmet: Türkiye Demiryolları Tarihi (1860–1953), İstanbul, 1953.

OPPENHEIM, Max Freiherr von: Die Beduinen, Band II: Die Beduinenstamme in Palaestina, Transjordanien, Sinai, Hedjaz, unter Mitarbeitung von E. Braunlich und W Caskel, Leipzig, 1943.

OPPENHEIM, Max Freiherr von: Vom Mittelmeer zum Persischen Golf, durch den Hauran, die syrische Wüste und Mesopotamien, Band II, Berlin, 1900.

OPPENHEIM, Max Freiherr von: Zur Entwicklung des Bagdadbahngebietes und insbesondere Syriens und Mesopotamiens unter Nutzanwenderung amerikanischer Erfahrungen, Berlin, 1904.

«Orientalia», Die Zukunft, Nr. 47, 22.8.1908.

ORTAYLI, İlber: İkinci Abdülhamit Döneminde Osmanlı İmparatorluğu'nda Alman Nüfuzu, Ankara, 1981.

ORTAYLI, İlber: İmparatorluğun En Uzun Yüzyılı, İstanbul, 1983.

ORTAYLI, İlber: "Osmanlı İmparatorluğu'nda Arap Milliyetçiliği", Tanzimat'tan Cumhuriyet'e Türkiye Ansiklopedisi, vol. 4, İstanbul, 1985.

ORTAYLI, İlber: "19. Yüzyılda Panislamizm ve Osmanlı Hilafeti", Türkiye Günlüğü, no. 31 (November–December, 1994).

ORTAYLI, İlber: "19. Yüzyıl Sonunda Suriye ve Lübnan Üzerinde Bazı, Notlar", Osmanlı Araştırmaları, vol. 4, İstanbul, 1984.

ORTAYLI, İlber: "Devenin Taşıma Eğrisi Üzerine Bir Deneme", Ankara Üniversitesi Siyasal Bilgiler Fakültesi Dergisi, vol. 28, no. 1–2 (March–June 1973).

ÖKE, Mim Kemal: İngiliz Casusu Prof. Ariminius Vambery'nin Gizli Raporlarında II. Abdülhamid ve Dönemi, İstanbul, 1983.

ÖKÇÜN, Gündüz: «Osmanlı Meclis-i Mebusanında Bağdat Demiryolu İmtiyazı Üzerine Yapılan Tartışmalar», Ankara Üniversitesi Siyasal Bilgiler Fakültesi Dergisi, XXV, 2 (June 1979).

ÖNSOY, Rıfat: Türk-Alman İktisadi Münasebetleri, (1871–1914), İstanbul, 1982.

ÖZCAN, Azmi: Pan – İslamizm, Osmanlı Devleti, Hindistan Müslümanları ve İngiltere (1877–1914), İstanbul, 1992.

ÖZTÜRK, Cemil: "Tanzimat Devrinde Bir Devletçilik Teşebbüsü: Haydarpaşa – İzmit Demiryolu", ÇYO, ed. E. İhsanoğlu, M. Kaçar, İstanbul, 1995.

ÖZYÜKSEL, Murat: Osmanlı-Alman İlişkilerinin Gelişim Sürecinde Anadolu ve Bağdat Demiryolları, İstanbul, 1988.

ÖZYÜKSEL, Murat: "Osmanlı Demiryolu İşletmeciliğinde Bir Devlet Girişimi: Hicaz Demiryolu" (I) and (II), İktisat Dergisi, no. 292, (March 1989) and no. 293 (April 1989).

ÖZYÜKSEL, Murat: "İktisadi Davranışlarımızın Tarihsel Kökenleri Üzerine Bir Deneme", İktisat Dergisi, no. 380, June/July, 1998.

ÖZYÜKSEL, Murat: "Bir Nüfuz Mücadelesi Aracı Olarak Suriye Demiryolları", Toplum ve Ekonomi, no. 10 (July 1997).

ÖZYÜKSEL, Murat: "Emperyalizm, Doğu Sorunu ve Osmanlı Demiryolları", İktisat Dergisi, no. 394 (October, 1999).

ÖZYÜKSEL, Murat: "Abdülhamit Dönemi Dış İlişkileri", Türk Dış Politikasının Analizi, ed. Faruk Sönmezoğlu, İstanbul, 1998.

ÖZYÜKSEL, Murat: Hicaz Demiryolu, İstanbul, 2000.

ÖZYÜKSEL, Murat: "Anatolian and Bagdad Railways", **The Great Ottoman Turkish Civilization**, vol. II, Ankara, 2002.

ÖZYÜKSEL, Murat: "Rail and Rule – Railway Building and Railway Politics in the Ottoman Empire", **Comparing Empires**, Jörn Leonhard and Ulrike von Hirschhausen (eds), Göttingen, 2011.

ÖZYÜKSEL, Murat: "Abdülhamid Devrinde Osmanlı-Alman İlişkilerinin Gelişimi", **Devr-i Hamid Sultan II. Abdülhamid**, Vol. 3, ed. M. Metin Hülagü, Ş. Batmaz, G. Alan, Kayseri, 2011.

ÖZYÜKSEL, Murat: **Osmanlı İmparatorluğu'nda Nüfuz Mücadelesi, Anadolu ve Bağdat Demiryolları**, İstanbul, 2013.

PAKALIN, Mehmet Zeki: " Surre Alayı", **Osmanlı Tarih Deyimleri ve Terimleri Sözlüğü**, vol. III, İstanbul 1993.

PAMUK, Şevket: «Osmanlı İmparatorluğu'nda Yabancı Sermaye: Sektörlere ve Sermayeyi İhraç Eden Ülkelere Göre Dağılımı (1854–1914)», **ODTÜ Gelişme Dergisi**, 1978 Special Edition.

PASCHASIUS: «Die Hedschasbahn», **AfEW**, 1927.

PEARS, Edwin: **Life of Abdul Hamid**, New York 1917.

PHILIPP, Hans Jürgen: «Der beduinische Widerstand gegen die Hedschasbahn», **Die Welt des Islams**, Band XXV (1985).

PICK, Walter: "Der deutsche Pionier Heinrich August Meissner Pascha und seine Eisenbahnbauten im nahen Osten 1901–1917", **Jahrbuch des Instituts für Deutsche Geschichte**, 4. Band (1975).

PICK, Walter: "Meissner Pasha and the construction of railways in Palestine and neighboring countries", **Ottoman Palestine 1800–1914**, ed. Gad G. Gilbar, Leiden, 1990.

POMIANKOWSKI, Joseph: **Der Zusammenbruch des ottomanischen Reiches, Erinnerungen an die Türkei aus der Zeit des Weltkrieges**, Zürich-Leipzig-Wien, 1928.

PÖNICKE, Herbert: **Die Hedschasbahn und Bagdadbahn, erbaut von Heinrich August Meissner Pascha**, Düsseldorf, 1958.

PRIATH, W.: «Die Eisenbahnen der asiatischen Türkei und ihre Bedentung in dem gegenwaertigen Kriege», **Verkehrstechnische Woche und Eisenbahntechnische Zeitschrift**, Nr. 34, 25.5.1915.

"Die projektierte Eisenbahn von Djeddah nach Mekka", **Frankfurter Zeitung**, 28.10.1904.

"The Proposed Damaskus Railway, Subscriptions from India", **Civil and Military Gazette**, 17.5.1904.

QUATAERT, Donald: **Social Disintegration and Popular Resistance in the Ottoman Empire, 1881–1908: Reactions to European Economic Penetration**, New York and London, 1983.

RATHMANN, Lothar: «Zur Legende vom antikolanialen Charakter der Bagdadbahnpolitik in der wilhelminischen Aera des deutschen Monopolkapitalismus», **Zeitschrift für Geschichtswissenschaft**, Sonderheft IX (1961).

RATHMANN, Lothar: **Stossrichtung Nahost 1914–1918**, Berlin, 1963.

RATHMANN, Lothar: **Berlin-Bağdat, Alman Emperyalizminin Türkiye'ye Girişi**, ed. Ragıp Zarakolu, İstanbul, 1982.

RAYMOND, Andre: **Osmanlı Döneminde Arap Kentleri**, İstanbul, 1995.

RAYMOND, Andre: "Arap Siyasal Sınırları İçinde Osmanlı Mirası", İmparatorluk Mirası, Balkanlar'da ve Ortadoğu'da Osmanlı Damgası, ed. L. Carl Brown, İstanbul, 2000.

RAYMOND, Andre: Arab cities in the Ottoman periode: Cairo, Syria, and the Maghreb, Burlington, 2002.

REIBEL, Willy: Die Gründung auslaendischer Eisenbahnunternehmungen durch deutsche Banken, Düsseldorf, 1934.

REINHARD, Rudolf: «Die Hedschasbahn», Universum, Moderne Illustrierte Wochenschrift, Nr. 5, Band 2, 1909.

RICHTER, Jan Btefan: Die Orientreise Kaiser Wilhelms II 1898, Eine Studie zur deutschen Aussenpolitik an der Wende zum 20. Jahrhundert, Hamburg, 1996.

ROHRBACH, Paul: "Bagdadbahn und Mekkabahn", Übersee, Nr. 4, 26.1.1908.

ROHRBACH, Paul: Die Bagdadbahn, Berlin 1911.

ROLF, H.: «Die erhöhte Bedeutung von Damaskus für den deutschen Export», DLZ, 15.7.1914.

ROLOFF, Max Breslau: Arabien und seine Bedeutung für die Erstaerkung des Osmanenreiches, Leipzig, 1915.

ROTHSTEIN, Th.: «Der Streit um die Bagdadbahn», Die Neue Zeit, Vol II, (1912/1913).

RUPPIN, A.: Syrien als Wirtschaftsgebiet, Berlin, 1917.

SAAD: "Die Mekkabahn und die Stadt Haifa am Karmelgebirge", Petermanns Mitteilungen, Nr. 51, Gotha, 1905.

SANDERS, Liman von: Türkiye'de 5 Yıl, İstanbul, 1968.

SARIYILDIZ, Gülden: «Hicaz'da Salgın Hastalıklar ve Osmanlı Devleti'nin Aldığı Bazı Önlemler», Tarih ve Toplum, no. 104 (August, 1992).

SARIYILDIZ, Gülden: Hicaz Karantina Teşkilatı 1865–1914, Ankara, 1996.

SCHAEFER, Carl Anton: Ziele und Wege für die jungtürkische Wirtschaft-spolitik, Karlsruhe, 1913.

SCHEKIB ARSLAN: "Um die Hedschaslinie – Die Kaempfe im Jordantal, um Amman und Es-Salt", Weser-Zeitung, Nr. 343 (15.5.1918).

SCHICK, Ernst: "Die Hedschasbahn» Technisches Magazin, no. 6 (1910).

SCHLAGINTWEIT, Max: «Die Mekkabahn», Asien, Heft 8 (1902).

SCHLAGINTWEIT, Max: Reise in Kleinasien, Münich, 1898.

SCHLAGINTWEIT, Max: Verkehrswege und Verkehrsprojekte in Vorder-asien, Berlin, 1906.

SCHMIDT, Hermann: Das Eisenbahnwesen in der asiatischen Türkei, Berlin, 1914.

SIRMA, İhsan Süreyya: Osmanlı Devleti'nin Yıkılışında Yemen İsyanları, İstanbul, 1994.

SIRMA, İhsan Süreyya: "Ondokuzuncu Yüzyıl Osmanlı Siyasetinde Büyük Rol Oynayan Tarikatlara Dair Bir Vesika", İstanbul Üniversitesi, Edebiyat Fakültesi Tarih Dergisi, no. 31, March 1977.

SIRMA, İhsan Süreyya: "Fransa'nın Kuzey Afrika'daki Sömürgeciliğine Karşı Sultan II. Abdülhamid'in Panislâmist Faaliyetlerine Ait Bir Kaç Vesika", İstanbul Üniversitesi Edebiyat Fakültesi Tarih Enstitüsü Dergisi, no. 7–8 (1977).

SINGER, Amy: Kadılar, Kullar, Kudüslü Köylüler, İstanbul 1996.

288   THE HEJAZ RAILWAY AND THE OTTOMAN EMPIRE

SİVRİKAYA, İbrahim: "Osmanlı İmparatorluğu İdaresindeki Aşiretlerin Eğitimi ve İlk Aşiret Mektebi", **Belgelerle Türk Tarihi Dergisi,** vol. XI, no. 63 (1972).

SLEMMAN, H.: «Le Chemin de fer de Damas – La Mecque», **Revue de l'Orient Chretien,** Year 5 (1900).

SÖLCH Werner: **Expresszüge im Vorderen Orient,** Düsseldorf, 1989.

STEPHENSON, MacDonald: **Railways in Turkey,** London, 1859.

STODDARD, Philip H.: **Teşkilât-ı Mahsusa,** İstanbul, 1993.

STUHLMANN, Franz: **Der Kampf um Arabien zwischen der Türkei und England,** Hamburg, Braunschweig, Berlin, 1916.

SULTAN ABDÜLHAMİT: **Siyasi Hatıratım,** İstanbul, 1984.

«Die syrischen Eisenbahnen Damaskus – Hamah und die Hedschasbahn», AfEW, (1905).

TAHSİN, Paşa: **Tahsin Paşa'nın Yıldız Hatıraları,** Sultan Abdülhamid, İstanbul, 1990.

TALAY, Aydın: **Eserleri ve Hizmetleriyle Sultan Abdülhamid,** İstanbul, 1991.

TEKİNDAĞ, Şehâbeddin: "Lübnan", **İslâm Ansiklopedisi,** vol. 7, İstanbul, 1972.

"The Sultan's First Railway Journey", **The Times,** 27.8.1867.

THIESS, F.: "Die Hedschasbahn", **Illustrierte Zeitung,** Nr. 3630, 23.1.1913.

TIBI, Bessam: **Arap Milliyetçiliği,** İstanbul, 1998.

TİMUR, Taner: "1838'de Türkiye, İngiltere Rusya", **Yapıt,** no. 10, (April/May, 1985).

TİMUR, Taner: "Osmanlı ve Batılılaşma", **Osmanlı Çalışmaları,** Ankara, 1989.

TOYDEMİR, Sait: "Hicaz Demiryolu İnşaatı Tarihinden", **Demiryollar Dergisi,** no. 275–278, 1948.

TOYNBEE, Arnold J.: **1920'lerde Türkiye, Hilafetin İlgası,** İstanbul, 1998.

TRUMPENER, Ulrich: "Almanya ve Osmanlı İmparatorluğu'nun Sonu", **Osmanlı İmparatorluğu'nun Sonu ve Büyük Güçler,** ed. Marian Kent, İstanbul, 1999.

«Die türkischen Eisenbahnen im Kriege», ZdVDEV, Nr. 12 (11 February 1914).

TÜRKÖNE, Mümtaz'er: **Siyasi İdeoloji Olarak İslamcılığın Doğuşu,** İstanbul, 1991.

TÜRKÖNE, Mümtaz'er: Özdağ Ümit (ed.): **Siyasi İslam ve Panislamizm,** Ankara, 1993.

UÇAROL, Rıfat: **Gazi Ahmet Muhtar Paşa, Askeri ve Siyasi Hayatı,** İstanbul, 1989.

UHLIG, Paul: **Deutsche Arbeit in Kleinasien von 1883 bis 1918,** Greifswald, 1925.

ULMAN, Haluk: **Birinci Dünya Savaşı'na Giden Yol,** Ankara, 1973.

ULMAN, Haluk: **1860–61 Suriye Buhranı, Osmanlı Diplomasisinden Bir Örnek Olay,** Ankara, 1966.

UZUNÇARŞILI, İsmail Hakkı: "Hicaz Vali ve Kumandanı Osman Nuri Paşa'nın Uydurma Bir İrade İle Mekke Emiri Şerif Abdülmuttalib'i Azletmesi" (offprint from **Belleten** no. 39), Ankara, 1946.

UZUNÇARŞILI, İsmail Hakkı: **Mekke-i Mükerreme Emirleri,** Ankara, 1972.

VELAY, A. du: **Türkiye Maliye Tarihi,** Maliye Bakanlığı Tetkik Kurulu Neşriyatı, Nr. 178, Ankara, 1978.

"Die Verkehrsgeographische Bedeutung der Vierlaenderecke bei Akaba", AfEW (1943).

«Die Vollendung der Hedschasbahn», AfEW, 1912.

WALLACH, Jehuda L.: «Ein deutscher Gelehrter und Offizier aeussert sich im ersten Weltkrieg zur Frage des militaerischen Wertes der Araberstaemme», Jahrbuch des Instituts für Deutsche Geschichte, vol. 4, 1975.

WALTER, Pick: «Der deutsche Pionier Heinrich August Meissner Pascha und seine Eisenbahnbauten im nahen Osten, 1901–1917», Jahrbuch des Instituts für Deutsche Geschichte, 1975.

WASTI, Syed Tanvir: "Muhammad Inshaullah and the Hijaz Railway", Middle East Journel (Spring 1998).

WAVELL, A.J.B., F.R.G.S.: A Modern Pilgrim in Mecca and a Siege in Sanaa, London, 1912.

WENSINCK, A. J.: "Hacc", İslâm Ansiklopedisi, vols 5/1, İstanbul, 1988.

WHIGHAM, H. J.: The Persian Problem, An Examination of the Rival Positions of Russia and Great Britain in Persia with some Account of the Persian Gulf and the Bagdad Railway, London, 1903.

WIEDEMANN, Max: Bagdad und Teheran; Politische Betrachtungen und Berichte, Deutsche Orient-Korrespondenz im Komission beim deutschen Kolonial-Verlag, Berlin, 1911.

WILSON, Arnold T.: The Persian Gulf, London, 1954.

WIRTH, Eugen: "Die Rolle tscherkessischer 'Wehrbauern' bei der Wiederbesiedlung von Steppe und Ödland im Osmanischen Reich", Bustan, Heft 1,1963.

YAVUZ, Hulusi: «Hicaz Demiryolu'nun Türkiye İktisadi ve Sosyal Tarihindeki Yeri ve Ehemmiyeti», V. Milletlerarası Türkiye Sosyal ve İktisat Tarihi Kongresi, Ankara, 1990.

YAVUZ, Hulusi: Osmanlı Devleti ve İslamiyet, İstanbul, 1991.

YENİAY, İ. Hakkı: Yeni Osmanlı Borçları Tarihi, İstanbul, 1964.

ZAIDI, Z. H.: "Chemin de fer du Hidjaz", Encyclopedie d' Islam, Tome II, Leiden, 1965.

ZANDER, Kurt: "Einwirkung der kleinasiatischen Eisenbahnen auf die Hebung des Grundbesitzes", AfEW, Berlin, 1894.

ZIFFER, E.A.: «Die mohammedanische Eisenbahn (Hedschasbahn)», Zeitschrift des österreichischen Ingenieur – und Architekten – Vereines, Nr. 9, 1910.

«Eine Zweigbahn von der Hedschasbahn nach der Aegyptischen Grenze, Zeitschrift des Vereins deutscher Ingenieure, vol. 59, Nr. 6, 6.2.1915.

IV- RESEARCHED NEWSPAPERS

Berliner Lokal Anzeiger, Berliner Tageblatt, Civil and Military Gazette, Daily News, The Daily Telegraph, Deutsche Levante Zeitung, The Evening Star, Financial News, Frankfurter Allgemeine Zeitung, Frankfurter Zeitung, Hamburgischer Correspondent, Hannoverscher Courier, The Globe, İkdam, Illustrierte Zeitung, Kölnische Zeitung, Manchester Guardian, The Morning Post, Münchner Neueste Nachrichten, Neue Freie Presse, Neue Preusische Zeitung, Neue Zürcher Zeitung, Orientalistische Litteratur-Zeitung, Die Post, Sabah, Schlesische Zeitung, Stamboul, The Times, Times of India, Technik und Wirtschaft, Die Umschau, Vossische Zeitung, Wiener Politische Correspondenz, Zeitung des Vereins Deutscher Eisenbahnverwaltungen, Die Zukunft.

INDEX